Missionaries of Science

MISSIONARIES OF SCIENCE

The Rockefeller Foundation and Latin America

Edited by
Marcos Cueto

Indiana University Press

Bloomington and Indianapolis

The paper used in this publication meets the minimum requirements of American
National Standard for Information Sciences—Permanence of Paper for Printed
Library Materials, ANSI Z39.48-1984.
⊗™
Manufactured in the United States of America

Library of Congress Cataloging-in-Publication Data

Missionaries of science : the Rockefeller Foundation and Latin America /
edited by Marcos Cueto.
p.   cm.—(Philanthropic studies)
Includes bibliographical references and index.
ISBN 0-253-31583-2 (cloth : alk. paper)
1. Science—Social aspects—Latin America.   2. Medicine—Social
aspects—Latin America.   3. Research—Latin America—International
cooperation.   4. Research—United States—International cooperation.
5. Medicine—Research—Latin America—International cooperation.
6. Medicine—Research—United States—International cooperation.
7. Rockefeller Foundation.   I. Cueto, Marcos.   II. Series.
Q175.L37M57   1994
306.4'5'0980904—dc20   93-24416

1 2 3 4 5 99 98 97 96 95 94

Contents

Preface

A FEW YEARS AGO, when I was beginning to work on my doctoral dissertation, I had an interview with the widow of a very important Peruvian
physiologist. As I explained my purpose of writing a social history of science
in Peru, she suddenly interrupted me. "My friend," she said, "I will tell you
the whole story; everything was done by a bunch of brilliant people and North
American money." I was struck by the response, not necessarily because it was
true but because it clearly reflected the common perception about the influential role played by American philanthropy in Latin American science and medicine.

My interest on this topic led me to visit, on several occasions, the Rockefeller Archive Center, and during 1990–91, I enjoyed a fruitful year as the
scholar-in-residence of the Center. As soon as I arrived, Darwin Stapleton, the
director of the Center, asked me to do "something else," in addition to my
research. The possibilities included participating in the processing of uncata-
logued manuscripts or the organization of one of the series of conferences regularly sponsored by the Rockefeller Archive Center on subjects germane to its
holdings.

I engaged myself, and the resources of the Center, in the organization of
a conference on a topic new to many: the history of the connections between
the Rockefeller Foundation and Latin America. Thanks to the cosponsorship
of the Rockefeller Archive Center and the Indiana University Center on Philanthropy, the conference took place on November 15, 1991, and was entitled,
"Science, Philanthropy and Latin America: Cross Cultural Encounters in the
Twentieth Century."

Five papers were presented by scholars who had spent a substantial
amount of time doing archival research at the Rockefeller Archive Center and
in archives of the Latin American countries of their expertise. The conference
ended with a final roundtable discussion by four outstanding scholars from
North and South America: Robert E. Kohler (University of Pennsylvania);
Lewis Pyenson (Université de Montréal); Nancy Leys Stepan (Columbia University) and Hebe Vessuri (Instituto Venezolano de Investigaciones Científicas). This volume publishes the proceedings of the conference, plus two additional essays that appeared a few years ago in a British journal and that share
topics and perspectives with the papers presented at the conference.

I would like to thank the speakers and commentators for their instructive and stimulating work, which made this volume possible. All the speakers are very grateful to the grant-in-aid program of the Rockefeller Archive Center, which made our research possible in the first place. One of the speakers, Joseph Cotter, suggested the final title of the volume. Thanks also to those who overcame distance and made special arrangements to participate in the discussion from the floor, including Paulo Gadelha, director of the Casa Oswaldo Cruz, who traveled from Rio de Janeiro, and Ann Emmanuelle Birn who came from Barcelona, and to Charles Godue and María Luisa Rodriguez, officers of the Panamerican Health Organization.

Thanks to David Edge and Sage Publications for giving permission to republish Deborah Fitzgerald's article, "Exporting American Agriculture: The Rockefeller Foundation in Mexico, 1943–1953," and my article, "The Rockefeller Foundation's Medical Policy and Scientific Research in Latin America: The Case of Physiology," which appeared in *Social Studies of Science* 16 (1986): 457–83, and 20 (1990): 229–54, respectively. I gratefully acknowledge Dwight F. Burlingame and Robert Fogal of the Indiana University Center on Philanthropy for their support; and the extraordinary efforts of Darwin Stapleton, Kenneth Rose, Lee R. Hiltzik, Gretchen S. Koerpel, Erwin Levold, Renee D. Mastrocco, Emily J. Oakhill, Harold Oakhill, Thomas Rosenbaum, Melissa A. Smith, Rosseann Variano, Penny Rieder, Camilla Harris, and Marie Calhahan of the Rockefeller Archive Center for helping so much to shape and organize the conference, and for providing the most delightful atmosphere for an academic discussion.

I would also like to thank Robert J. Sloan and the staff of Indiana University Press for their professional and kind editorial assistance.

Marcos Cueto

Introduction

THE ORIGINS OF this volume can be traced to the organization of a conference, "Science, Philanthropy and Latin America: Cross Cultural Encounters in the Twentieth Century," which took place in November of 1991 at the
Rockefeller Archive Center (RAC) in North Tarrytown, New York. The event
brought together a number of researchers who were studying the relationship
between U.S. philanthropy and Latin American science and medicine.[1] Their
work appeared novel because it intertwined the social history of Latin America
and the history of science, combining materials from U.S. and Latin American
archives, and paying special attention to the local reception of and responses
to U.S. philanthropy.

This last feature was a major departure from a number of critical, and non-
critical, studies of the history of U.S. philanthropy in less-developed countries.
For many years, such studies have concentrated on the motivation of the sending side, i.e., those U.S. institutions administering philanthropic support
abroad. The intentions attributed to the donors included: a new humanitarian
concern for poor countries, the protection of the productivity of the tropical
areas of the world under U.S. influence, the control of the native ruling elites
through the subtle mechanisms of cultural hegemony, and the use of philanthropy as an arm of imperial U.S. diplomacy.[2] In general, these works paid
scant attention to local historical actors, to the role played by the foundation's
field officers in the adaptation of programs, and to general questions of accommodation and negotiation of the changing policies of U.S. foundations in the
third world.

In an effort to present a more balanced and dynamic view of the relationship between foreign philanthropy and science, the chapters of this volume are
more attentive to developments on the receiving side, i.e., the institutions and
people that were the object of the philanthropic programs. Not all of the interpretations are identical, and this volume is by no means a definitive account
of the role played by the Rockefeller Foundation (RF) in Latin America. The
general aim is to illustrate the first years of the foundation's contact with Latin
America, to contrast the different emphases assigned to countries and RF programs, and, moreover, to indicate a future trend of scholarship by presenting
detailed case studies that might suggest elements for a general interpretation.
New studies that could contribute to this interpretation should appear, if more

researchers become aware of the unique opportunities for cross-cultural studies involved in the study of U.S. philanthropy abroad, and become familiar with the extraordinary quantity and quality of materials that exist in repositories such as the Rockefeller Archive Center.[3]

The RAC holds the papers of the RF, the oldest and, for much of the twentieth century, the most important American philanthropy working in Latin America.[4] Between 1917 and 1962, more than 1,700 Latin Americans received fellowships from the RF (see Table 1). During the initial years, the RF's medical activities in Latin America were concentrated in public health and medicine.[5] The John Simon Guggenheim Memorial Foundation, probably the second- or third-most-important American foundation supporting Latin American medicine and science, awarded a total of 610 fellowships to Latin American scholars of all fields between 1930 and 1967.[6]

Initially, RF activities in the region were carried out mainly through the International Health Commission (later Board and later Division), organized in 1913 with the goal of extending the work of the Rockefeller Sanitary Commission for the Eradication of Hookworm Disease to countries other than the United States.[7] The first campaign against hookworm in Latin America began in 1916 in Brazil, as a byproduct of a survey of the RF medical commission.[8] The Brazilian government created a Department of Uncinariasis and appointed as its director Lewis Hackett, an RF officer who had organized demonstrations and campaigns in selected areas of eleven states. An outgrowth of the hookworm campaign was the organization of rural sanitary services in a number of these states.[9] Additional surveys and campaigns on hookworm were developed by the RF in Colombia (1920–29), Argentina (1923–27), Paraguay (1922–27), Venezuela (1927–28), and several Central American and Caribbean countries. The tendency of RF work on hookworm in Latin America was to leave the domain of simple demonstration and to take on something of the attributes of a state public health service, with hope of promoting a reform of public health services.[10] Two goals that were not fully achieved—and their unfulfillment partially explains why hookworm rapidly lost its original priority position in RF's agenda—were the eradication of hookworm and the use of the hookworm campaigns to stimulate the reform of public health in Latin America.

Shortly after the RF's creation, William C. Gorgas, Surgeon General of the United States Army and director of the yellow-fever campaigns in Havana and Panama, was influential in reorienting the foundation's original goal. After 1916, the RF was convinced that the first worldwide sanitary task that might be undertaken with promise of lasting results was the eradication of yellow fever.[11] As a result, the International Health Commission decided to eliminate yellow fever from the Latin American cities in which the disease was endemic. Between 1923 and 1935, yellow-fever studies and control activities represented

Table 1. Number of Fellowships, Scholarships, and Training Awards
given to Latin Americans by the Rockefeller Foundation, 1917–1962

	Agricultural and Natural Sciences	Medical, Health, and Population Sciences*	Humanities and Social Sciences	Total
Argentina	39	77	11	127
Bolivia	9	6	1	16
Brazil	140	274	29	443
Chile	58	120	36	214
Colombia	100	138	26	264
Costa Rica	16	15	2	33
Cuba	1	12	3	16
Dominican Republic	0	5	0	5
Ecuador	11	16	1	28
El Salvador	0	21	0	21
Guatemala	9	14	0	23
Honduras	7	0	0	7
Mexico	198	114	47	359
Nicaragua	5	15	0	20
Peru	50	34	5	89
Puerto Rico	0	2	12	14
Uruguay	16	12	2	30
Venezuela	5	5	2	12
Total	664	880	177	1,721

*Includes the International Health Division and the Medical and Natural Sciences Division.
Source: Latin American Awards, 7-17-1963, RFA, RG 1.2, Series 300, Box 2, Folder 8, RAC.

16 percent (U.S. $4.8 million) of the total operating cost of the RF, and between
1936 and 1944, yellow-fever expenditures continued at a high level, namely,
about $2 million for the nine years, the largest fraction of which was spent in
South America.[12] Between 1913 and 1940, the foundation spent a little more
than U.S. $13 million in Latin America, of which almost 50 percent was used
in the study and control of yellow fever.[13] The reasons that explain the RF's
concentration on yellow fever include: the protection of international commerce, the fear of reinfection of the United States, the possibility of demonstrating rapid success, the power of scientific communities in defining philanthropic priorities, and the needs and responses of some emergent Latin
American nation-states. The RF's program in yellow fever was not always taking place in areas where Standard Oil had investments, and can be better understood as part of a concern for the stability of the international capitalist system as a whole, and for the cultural and political role that the United States
was beginning to play in it.

The opening of the Panama Canal in 1914 created the possibility of the

Caribbean exportation of yellow fever into tropical Asia, a region until then free of the disease. U.S. interests and the U.S. Army had had responsibilities in these regions since the Spanish-American War and the occupation of the Philippines at the turn of the twentieth century. Yellow fever not only posed a threat for American colonists living outside the United States but also carried the potential to reinfect the southern United States, a subtropical region that had suffered severe yellow-fever epidemics in the nineteenth century. It was this concern for the national security of the United States that partially explains the RF's emphasis on yellow fever.

During the early twentieth century, the eradication of yellow fever from big urban centers seemed to many experts a clear possibility. Methods for its control were successfully defined in 1900, when a U.S. Army commission, working in Havana, investigated the theory concerning transmission of yellow fever by the *Aedes aegypti* mosquito.[14] Nevertheless, even in 1916 when the RF sent a commission to Latin America to survey the yellow-fever situation, several crucial aspects of the disease were not understood. The microorganism involved in the disease was unknown, and this ignorance led to a race to determine the yellow-fever causative agent among U.S., Latin American, and European scientists.

U.S. scientists played an important role in reinforcing the idea that yellow fever was a major international problem. The story of the foundation, particularly the story of its public health work, is linked to the first of the large philanthropies of John D. Rockefeller—the Rockefeller Institute of Medical Research—located in New York City. Scientists from this institute focused great attention on resolving the scientific mysteries of yellow fever.[15] Beginning in 1928, the International Health Division of the RF operated a yellow-fever laboratory at the Rockefeller Institute. The study of the disease was one of the first areas in which U.S. scientists working in tropical medicine achieved international recognition, surpassing European scientists who had an early lead in the field.

The interest in yellow fever led to RF campaigns in Ecuador (1918); Guatemala (1919); Peru (1920–22); Central America (1923); Mexico (1923); and Brazil, where a yellow-fever service was under the control of the RF for sixteen years (1923–40), an important exception to the foundation's policy of avoiding active participation in the administration of government services for a prolonged period of time. These campaigns were instrumental in the disappearance of urban and maritime yellow fever from the Americas and to the eradication of the vector *Aedes aegypti* from many Brazilian and South American cities. However, the discovery of jungle yellow fever in South America between 1928 and 1934 demonstrated that yellow fever was transmitted by mosquitoes other than the *Aedes*, through monkeys and other animals, and that there was a permanent reservoir of infection in the jungles.[16] During the 1930s, the foundation began to lose interest in yellow fever, partially because the goal of eradication became infeasible.

In addition to hookworm and yellow fever, malaria attracted the attention of the RF. The major campaign against this disease occurred in Brazil in 1938, when the African *Anopheles gambiae*, the most dangerous mosquito of its family, was responsible for the greatest epidemic of malaria ever to hit the Americas. The epidemic encompassed a circle of a two-hundred-mile radius around Natal in the Jaguaribe valley in the state of Ceara, producing 100,000 cases and at least 14,000 deaths. A special antimalaria service was created and placed under the direction of Fred L. Soper, then the chief foundation representative in Brazil. In a short period, Soper organized a staff that, at its peak of activity, had some 4,000 employees.[17] The strategy of the anti-*gambiae* campaign was to destroy all larvae and adult mosquitoes in the infested territories and adjacent areas, to disinsecticize all vehicles leaving the *gambiae* zone, and by systematic surveys, gauge the results of the operations. By the end of 1940, malaria was infrequent in the previously infected area, and no trace of the insect was reported in Brazil.[18] The eradication of *gambiae* in Brazil was instrumental in rehabilitating the eradication concept after the failed experiences with hookworm and yellow fever and in stimulating RF campaigns against malaria in Sardinia and upper Egypt.

Another factor involved in explaining the public health program of the RF in the region is the response of the Latin American governments and the local medical communities: they were willing to accept American aid because they lacked a nationally organized public health bureaucratic apparatus. RF aid to the organization of such an apparatus was welcomed in a period when many of these governments were trying to expand federal control over provincial authorities in a number of areas, including sanitation.[19] Latin American governments were also interested in improving public health in their ports and cities as a means of avoiding the effects of quarantine and of attracting highly desired European immigration.

During the 1920s, the RF began to send staff members on long-term assignments to reside in Latin America. As early as 1926, thirty-seven foundation staff members were stationed in fourteen Latin American countries.[20] These assignments enhanced the importance of the RF because, until the early 1940s, there was no effective international framework through which Latin American countries could act on common health problems. The Pan American Sanitary Bureau, created in 1902, functioned until the late 1930s with a small staff as a virtual branch of the U.S. Public Health Service.[21]

After a major reorganization in 1928, the RF continued its emphasis on public health and medicine but began to pay more attention to scientific education. Such aid was given under the assumption that the health of populations depended not on the RF's campaigns but on the quality of the training of local physicians. This was when the RF awarded its first grants for Latin American life scientists, especially those working on physiology and neurophysiology in medical schools.

The Second World War was a turning point for the relationship between the RF and Latin America. After the suspension of the RF's European activities during the 1940s, new possibilities for U.S. philanthropic work in Latin America appeared. After 1940, the foundation began to alter its public health emphasis to promote agriculture, under the assumption that a "traditional" agricultural sector and poor nutrition were the main factors retarding Latin American development.

Between 1940 and 1949, the foundation launched a major agricultural program in Mexico, investing more than U.S. $12.5 million in that decade.[22] The program had two main goals: to improve the efficiency of Mexican food-crop production, and to train Mexicans in agricultural techniques. The major work of the foundation, concentrating on genetic improvements in Mexican corn and wheat, was centered at an experimental station located at the National College of Agriculture. The program conducted tests of new fertilizers and hybrid varieties, and provided Mexicans with fellowships for study in the United States. In 1950 and in 1955, the RF set up similar agricultural programs in Colombia and Chile.

The support of agriculture signified an expansion beyond the RF's early attention to medical work, which continued until 1951. The RF's public health activities declined worldwide in that year when it merged the International Health Division with the medical science program. One of the new areas that the RF supported was the development of genetics, partly connected with agricultural studies, in selected universities of Latin America. During the same period, the foundation became interested in supporting a number of Latin American institutions of higher learning. The most remarkable of these were the Instituto Tecnológico de Monterrey in Mexico, the University of Sao Paulo at Ribeirao Preto, and the Universidad del Valle in Cali, Colombia—new, provincial universities located in or near booming industrial areas. These conditions were perceived as potential advantages with respect to avoiding the resistance of more traditional institutions and finding local resources for the support of first-rate educational institutions.

In the early 1960s, the RF was still among the most-active American agencies that financed Latin American science, giving U.S. $8.3 million between 1960 and 1962 for science and technology projects. The amount represented more than half of what the U.S. government awarded in the same areas, and 36 percent of what private American foundations donated to Latin American science and technology.[23]

As this brief overview suggests, the Rockefeller Foundation played a crucial role in several sectors of twentieth-century Latin American science and society. The materials at the RAC cover the development of this role in Latin American medicine, public health, science, agriculture, the social sciences, and the humanities from the 1910s to the early 1970s. Moreover, these materials illustrate how, in many of these areas, there was an increasing process of "Americanization" and a corresponding decline of European influence, especially French influence, in Latin American medicine and medical sciences.

The majority of the Latin American materials at the RAC are kept in the collection entitled Rockefeller Foundation Archives, which includes the documents of the International Health Division and the field offices located in several Latin American cities. As an example, just one record group of this collection, (Record Group 1) called "Projects," includes 443 boxes of Latin American materials, of which 165 relate to grants offered in medicine and public health, 219 to natural sciences and agriculture grants, and 65 to grants in the social sciences and humanities. In this record group, materials related to grants to Brazilian institutions are kept in 122 boxes, grants to Colombian institutions in 105 boxes, and grants to Mexican institutions in 99 boxes. Another record group is devoted to the 50 boxes of the recently processed records of the Mexico Field Office, originally established in 1943.[24] Finally, the papers of the various RF officers who worked in Latin America are maintained as a separate, special collection and include 45 boxes of the papers of George Harrar, the first director of the Mexico Field Office, later president of the Rockefeller Foundation, and considered the father of the Green Revolution. Among the many Latin American topics very well represented in the RAC collections are: the history of yellow fever in the twentieth century, campaigns against hookworm, the developments of the basic sciences in the schools of medicine, agricultural development, the use of DDT against malaria, the beginnings of professional nursing, and the first schools of social sciences in Latin America.[25]

Some of these topics were covered by the papers presented at the conference, which were prepared especially for the event, although some are parts of larger projects that the authors have in progress. Special attention was given to Mexico and Brazil, countries that received sizable donations from the Rockefeller Foundation. Two of the chapters in this volume were not originally papers presented at the conference, Deborah Fitzgerald's "Exporting American Agriculture," and my article on "The Rockefeller Foundation's Medical Policy." They appeared in a British journal that probably few Latin American historians or scholars of philanthropy read, and their inclusion here contributes to a more comprehensive volume on the activities of the RF in the region.[26]

My chapter, "Visions of Science and Development: The Rockefeller Foundation's Latin American Surveys of the 1920s," analyzes the surveys the RF made during the 1920s on the medical, scientific, and public health conditions of fifteen Latin American countries. These surveys provide valuable information on Latin American education and research, and a unique opportunity to observe how institutions and cultural traditions from a different society were perceived and portrayed. The surveys were used to elaborate the policy and programs of the foundation in a given country.

Steven C. Williams's "Nationalism and Public Health: The Convergence of Rockefeller Foundation Technique and Brazilian Federal Authority during the Time of Yellow Fever, 1925–1930" deals with the aspirations on the part of the RF to create a unified anti-yellow-fever system in Rio de Janeiro, which resulted in some confrontations over methods, personnel, and allocation of resources that hindered Rockefeller efforts. This work underscores the pragmatic political approach that some Rockefeller officials undertook as a way to achieve a coordinated, unified anti-yellow-fever program in Brazil.

Another chapter on yellow fever, Armando Solorzano's "The Rockefeller Foundation in Revolutionary Mexico: Yellow Fever in Yucatan and Veracruz," shows that the dissimilar economic and political conditions in two provinces of revolutionary Mexico—Yucatan and Veracruz—changed in a substantial manner the objectives and techniques employed in the RF public health campaigns. In Veracruz in the 1920s, for example, the RF campaign emphasized interaction with the people and acted as an important component in reversing the anti-U.S. sentiments of the population.

Two studies deal with one of the main chapters of American philanthropy abroad: the Green Revolution in Mexico. Deborah Fitzgerald's "Exporting American Agriculture: The Rockefeller Foundation in Mexico, 1943–1953" offers insight into the tension and conflict present in the transplantation of an American model to foreign conditions. According to Fitzgerald, the agricultural program succeeded where crops and farmers had a close resemblance to American crop farmers, and failed where the American model did not fit local conditions. From a different point of view, Joseph Cotter's "The Rockefeller Foundation's Mexican Agricultural Project: A Cross-Cultural Encounter, 1943–1949" traces the early success of the foundation's programs in Mexico and argues that the Mexican agricultural research community was eager to accept U.S. aid because of more than twenty years of professional failure and crisis. As a result, the values of the recipient merged with those of the foundation, at least for some years.

My chapter on Latin American physiology and Thomas Glick's on Brazilian genetics follow the emergence and growth of Latin American scientific disciplines, assess the effect and limitations of philanthropic scientific policy, and illustrate the interaction between U.S. and Latin American scientists eagerly promoted by the RF. "The Rockefeller Foundation's Medical Policy and Scientific Research in Latin America: The Case of Physiology" describes the changes in foundation policy during the 1940s and the relationship between philanthropic policies, medical reform, and institution-building efforts in physiology for four Latin American countries. "The Rockefeller Foundation and the Emergence of Genetics in Brazil, 1943–1960" studies an important area of the foundation's activity: support of basic research in the life sciences, which was particularly intense in Brazil. This chapter traces the emergence of an international network of research under the influence of the distinguished scientist Theodosius Dobzhanksy, and discusses how the institutional setup of the Brazilian laboratories supported by the Rockefeller Foundation was used to channel a particular kind of genetics research.

The title of the volume provides a general unifying theme for the chapters. In the broadest sense, the RF can be traced to earlier efforts of Protestant congregations in Latin America, which were very active in promoting education and public health in some Latin American countries around the turn of the twentieth century. The RF could be perceived also as a missionary organization, in the sense that it initially went to Latin America with a set of values that it believed would modernize a "traditional" culture. Among these values, the foundation considered those embodied in "Western science" and the institutional enactment of the American university as the main tools for the modernization of "backward" and non-core Western cultures. In several cases, these values conflicted with those of the recipient culture, producing conflict or accommodation, and raising the question of who controlled the direction, mode, and pace of the process of change. In other cases, the needs of the recipient merged with some of the modernizing goals of the foundation. The chapters of this volume analyze the diverse reactions to philanthropic policies encouraged from abroad, offering some new answers to old questions, and, I hope, suggesting new questions on the role played by U.S. philanthropy in Latin America and other parts of the third world.

Notes

1. There are a number of studies on Latin America that use the Rockefeller Foundation's materials. See Guillermo Arbona and Annete B. Ramirez de Arellano, *Regionalization of Health Services: The Puerto Rican Experience* (Oxford: Oxford University Press, 1978); Kenneth F. Kiple, *The Caribbean Slave: A Biological History* (Cambridge: Cambridge University Press, 1984); Annete B. Ramirez de Arellano and Conrad Seipp, *Colonialism, Catholicism, and Contraception: A History of Birth Control in Puerto Rico* (Chapel Hill: University of North Carolina Press, 1983); Annete B. Ramirez de Arellano, "The Politics of Public Health in Puerto Rico, 1926–1940," *Revista de Salud Pública de Puerto Rico* 3 (1981): 35–73; Marcos Cueto, *Excelencia*

Científica en la Periferia, Actividades Científicas e Investigación Biomédica en el Perú, 1890–1950 (Lima: Tarea, 1989); Marcos Cueto, "The Rockefeller Foundation's Medical Policy and Scientific Research in Latin America: The Case of Physiology," *Social Studies of Science* 20 (1990): 229–54; Marcos Cueto, "Andean Biology in Peru: Scientific Styles in the Periphery," *Isis* 80 (1989): 640–58; Marcos Cueto, "Sanitation from Above: Yellow Fever and Foreign Intervention in Perú, 1919–1922," *Hispanic American Historical Review* 72 (February 1992): 1–22; Armando Solórzano, "The Rockefeller Foundation in Mexico: Nationalism, Public Health and Yellow Fever, 1911–1924" (Ph.D. dissertation, University of Wisconsin, Madison, 1990); and Deborah Fitzgerald, "Exporting American Agriculture: The Rockefeller Foundation in Mexico, 1943–53," *Social Studies of Science* 16 (1986): 457–83.

2. These pioneering studies are an important point of reference but today are considered as too simplistic: Robert Arnove, *Philanthropy and Cultural Imperialism: The Foundations at Home and Abroad* (Boston: G. K. Hall, 1980); Edward H. Berman, *The Influence of the Carnegie, Ford, and Rockefeller Foundations on American Foreign Policy* (Albany: State University of New York Press, 1983); E. Richard Brown, "Public Health and Imperialism: Early Rockefeller Programs at Home and Abroad," *American Journal of Public Health* 66 (September 1976): 897–903; Merle Curti, *American Philanthropy Abroad* (New Brunswick, N.J.: Transaction Books, 1988); and Saul Franco Agudelo, "The Rockefeller Foundation's Antimalarial Program in Latin America: Donating or Dominating?" *International Journal of Health Services* 1 (1983): 51–57.

3. For other collections on American philanthropy that might be of interest, see Kenneth W. Rose, *The Availability of Foundation Records: A Guide for Researchers* (New York: Rockefeller Archive Center, 1991).

4. A general description of the Latin American materials at the Rockefeller Archive Center appears in Marcos Cueto, "El Rockefeller Archive Center y la Medicina, la Ciencia y la Agricultura Latinoamericanas del Siglo Veinte: Una Revisión de Fondos Documentales," *Quipu: Revista Latinoamericana de Historia de la Ciencia y la Tecnología* 8 (1991): 35–50. The Rockefeller Archive Center also keeps the papers of the Rockefeller family, of the various philanthropic institutions that members of the family created, and of other philanthropic institutions not connected with the family, such as the Commonwealth Fund and the Russell Sage Foundation. The general guide to the RAC is Emily J. Oakhill and Kenneth W. Rose, *A Guide to Archives and Manuscripts at the Rockefeller Archive Center* (New York: Rockefeller Archive Center, 1989). Collections that might be of interest to the Latin American economic historian are: American International Association for Economic and Social Development Archives, and International Basic Economic Corporation Archives, and the papers of Wayne G. Broehl, John R. Camp, John D. Rockefeller, Jr. and Nelson A. Rockefeller. A work done with some papers of these collections is Elizabeth A. Cobbs, "Entrepreneurship as Diplomacy: Nelson Rockefeller and the Development of the Brazilian Capital Market," *Business History Review* 63 (1989): 88–121.

5. For a general history of the Foundation, see Raymond B. Fosdick, *The Story of the Rockefeller Foundation* (New Brunswick, N.J.: Transaction Publishers, 1989); and Robert Shaplen, *Toward the Well-Being of Mankind: Fifty Years of the Rockefeller Foundation* (New York: Doubleday and Company, 1964).

6. Latin American Awards, July 17, 1963; and Rockefeller Foundation, "Expenditures for Work in Latin America," Rockefeller Foundation archives (hereafter RFA), R.G 1.2, Series 300, Box 2, Folder 8, Rockefeller Archive Center (hereafter RAC); and John Simon Guggenheim Memorial Foundation, *Directory of Fellows, 1925–1967* (New York: John Simon Guggenheim Memorial Foundation, 1968).

7. A history of this division appears in Lewis W. Hackett, "Once upon a Time," *American Journal of Tropical Medicine and Hygiene* 9 (1960): 105–15; and Greer Williams, *Plaque Killers* (New York: Scribners, 1969). The classic study of the work against hookworm in the southern United States is John Ettling, *The Germ of Laziness: Rockefeller Philanthropy and Public Health in the New South* (Cambridge, Mass.: Harvard University Press, 1981).

8. The visit followed the rule of the RF to develop programs only where local conditions

were known. "Medical Education in Brazil, Diaries, Memoranda, Notes, Reports by the Medical Commission to Brazil 1916, Lewis W. Hackett 1919, Dr. Robert A. Lambert 1923–25, George K. Strode 1925–26," RFA, Series 305, Box 2, Folder 15, RAC.

9. "Summary of International Health Division—South America," 38, RFA, R.G. 3.1, Series 908, Subseries Pro-11, Box 12, Folder 129, RAC.

10. Lewis Hackett, "Relief and Control of Hookworm Disease in Brazil, from November 22, 1916, to December 31, 1919," RFA, RG 5, Series 2, Box 24, Folder 148, RAC.

11. Wickliffe Rose, "Memorandum on Yellow Fever: Feasibility of Its Eradication, 7 October 1914," RFA, R.G. 5, International Health Board/Division, Series 300, Box 22, Folder 135, RAC.

12. Note concerning the table (R.C. 5) of expenditures of the I.H.D, 1913–1949, RFA, R.G. 3, Series 908, Box 14, Folder 145, RAC.

13. International Health Division, April 4, 1941, RFA, R.G. 1.2, Series 300, Box 1, Folder 7; and Rockefeller Foundation, "Appropriation to Latin America (All Divisions), 1913–1949," RFA, R.G. 1.2, Series 300, Box 2, Folder 9, RAC.

14. Nancy Stepan, "The Interplay between Socio-economic Factors and Medical Science: Yellow Fever Research, Cuba and the United States," *Social Studies of Science* 8 (1978): 397–424.

15. Between 1926 and 1936, they published 200 works on yellow fever. They appear in *Collected Papers on Yellow Fever by Members of the Staff of the International Health Division of the Rockefeller Foundation*, 4 vols. (New York: International Health Division of the Rockefeller Foundation, n.d.).

16. See Fred. L. Soper, "Yellow Fever: The Present Situation (October 1938) with Special Reference to South America," *Transactions of the Royal Society of Tropical Medicine and Hygiene* 32 (1938): 297–332.

17. Between 1938 and 1940, the RF appropriated a little more than U.S. $330,000 for control measures for malaria in Brazil. "Brazil *Anopheles Gambiae* Control," RFA, R.G. 1.1, Series 305, Box 16, Folder 14, RAC.

18. See Fred L. Soper and D. Bruce Wilson, "*Anopheles gambiae* in Brazil, 1930 to 1940," in J. Austin Kerr, ed., *Building the Health Bridge, Selections from the Works of Fred L. Soper* (Bloomington: Indiana University Press, 1970), 247–98.

19. See Luiz Antonio de Castro Santos, "Power, Ideology and Public Health in Brazil, 1889–1930," (Ph.D. dissertation: Harvard University, 1987).

20. The Rockefeller Foundation, *Annual Report, 1961* (New York: Rockefeller Foundation, 1961), 103.

21. The bureau was not really an organization but a board of changing composition, whose members continued to reside and be employed in their own countries. In 1920, Hugh Smith Cumming was appointed U.S. Surgeon General and also director of the bureau. He was repeatedly reelected and acted in this dual capacity until 1936. See N. Howard-Jones, "The Pan American Health Organization: Origins and Evolution," *WHO Chronicle* 34 (1980): 419–26.

22. For a general history of the program, see Bruce H. Jennings, *Foundations of International Agricultural Research: Science and Politics in Mexican Agriculture* (Boulder, Colo.: West View Press, 1988).

23. U.S. Department of State, Office of International Scientific Affairs, *Inventory of Grants for Science and Technology by U.S. Government Agencies, UNTA, Special Fund, Ford, Kellogg and Rockefeller to Latin American Countries, 1960–1962* (Washington, D.C.: U.S. Department of State, n.d.). The other private American foundations included in this survey were the Ford and the Kellogg foundations.

24. A description of the materials appears in Erwin Levold, "RF Mexico Field Office and the Green Revolution," *Rockefeller Archive Center Newsletter* (Summer 1989): 10–11; and Erwin Levold, "The Rockefeller Foundation's Mexico Field Office and the Central American Corn Improvement Project," *The Mendel Newsletter: Archival Resources for the History of Genetics and Allied Sciences* 1 (June 1991): 4–6. The first years of the Mexico Field Office and the begin-

nings of the Green Revolution in Mexico are narrated in E. C. Stakman, Richard Bradfield, and Paul C. Mangelsdorf, *Campaigns against Hunger* (Cambridge, Mass.: Harvard University Press, 1969). Other Latin American field offices whose materials are being processed are Belem and Rio de Janeiro in Brazil; Santiago in Chile; Bogotá in Colombia, and Guatemala.

25. There is a guide for the nursing materials: Emily J. Oakhill, *A Survey of Sources for the History of Nursing and Nursing Education at the Rockefeller Archive Center* (New York: Rockefeller Archive Center, 1990). There is also a guide to the extraordinary collection of photographs at the RAC: *Photograph Collections in the Rockefeller Archive Center* (New York: Rockefeller Archive Center, 1986).

26. A promising new area of research not included in this volume is the relationship between the Rockefeller Foundation and Latin American social sciences and humanities. See Servando Ortoll, "The RF, Mexican History and Spanish Refugee Scholars," *Rockefeller Archive Center Newsletter* (Summer 1989): 4–5; and Jesús Mendez, "Foreign Influences and Domestic Needs in Intellectual Institution Building: The Gestación of the Casa de España/Colegio de Mexico," *SECOLAS, Annals. Journal of the Southeastern Council on Latin American Studies* 21 (1990): 5–23.

1

Visions of Science and Development

The Rockefeller Foundation's Latin American Surveys of the 1920s

Marcos Cueto

S HORTLY AFTER THEIR creation in the early twentieth century, American philanthropies began to export to less-developed countries institutional models for the organization of knowledge, under the assumption that the giver knew what was good for the receiver. This assumption reflected the institutions' perceptions of isolated elements of foreign cultures, and their comparisons of these perceptions with social, cultural, and scientific developments in the United States. The perceptions are difficult to identify in the policies toward a specific laboratory or in the day-to-day administration of grants. However, the Rockefeller Foundation made frequent use of a program-planning tool that concentrated many of these perceptions: the surveys.

Between 1916 and 1929, officers of the RF surveyed the medical, scientific, and public health conditions in fifteen Latin American countries.[1] In conducting these surveys, the foundation expanded to an international scale the general practice followed by other early-twentieth-century American philanthropies, especially those of the so-called Progressive Era, when they conducted social surveys before initiating any action.[2] The RF surveys provide very valuable information on education and research in Latin America during the early twentieth century. They also provide a unique opportunity to observe how new realities were perceived, portrayed, and incorporated into philanthropic discourse, making them intelligible to U.S. audiences. The reports reveal the main assumptions about science, medicine, and development that informed the early philanthropic efforts in Latin America.

This inclination to survey continued to influence the way the foundation worked, and succeeding reports, such as those undertaken in the 1950s for medical education, preceded changes in foundation programs. The surveys became one of the main symbols of "scientific philanthropy," a never well de-fined method by which the RF claimed to make rational decisions. However, as this chapter tries to demonstrate, the surveys combined ethnocentric no-tions with an elitist reading of the scientific development of the United States.[3]

The Rockefeller Foundation's Initial Involvement in Latin America

In 1913 the International Health Commission (IHC) of the RF was organ-ized, with the goal of undertaking the international diffusion of the work done by the Rockefeller Sanitary Commission for the Eradication of Hookworm Dis-ease in the south of the United States.[4] The foundation decided to begin oper-ations in the British West Indies, and in other non-Caribbean, tropical colonies of Great Britain.[5] The West Indies islands offered the advantage of being small and geographically near the United States, affording the foundation opportu-nity to exercise close supervision at not too great expense. In addition, the is-lands attracted little attention, which permitted the foundation to carry out its international experiment in a "quiet corner" without committing the RF to a major program.[6] This careful start reflected a characteristic preference for cau-tion on the part of the Rockefeller philanthropic advisors. This cautious ap-proach may have stemmed from the history of criticism that the foundation's creator, John D. Rockefeller, had endured, as investigative journalists exam-ined the practices of big American corporations in the early twentieth century.

A second step in the RF's early involvement with Latin America occurred around 1914, when the foundation decided to work in Central America. In that region, an important pattern was set: work would be undertaken only when the approval and cooperation of local governments was secured.[7] In these and future movements of the foundation in Latin America, all major decisions were first discussed with the U.S. ambassadors to the respective republics, the De-partment of State in Washington, and the authorities of the prospective host nations. The Latin American governments welcomed the RF's activities be-cause these governments were in a process of expanding the legitimacy of the state to a national level.[8] This expansion included the creation of a national public health infrastructure, built with foundation aid.

Yellow fever gave rise to one of the first major interventions of the RF in Latin American sanitation. The disease carried the potential to reinfect the southern United States, which had suffered epidemics in the nineteenth cen-tury. The development of successful techniques to control the vector of the fever convinced the RF that the disease could be eradicated rapidly from the Americas.[9] In 1916, an RF yellow-fever commission visited several Latin American countries, and RF campaigns followed on the heels of the visit.[10] These campaigns supported the increased presence of U.S. interests in the region. Shortly before the First World War, North American capital flowed into Latin American countries, helping to consolidate an export economy based on raw materials.[11] The work of the RF served to maintain a stable world trade and to protect the health of U.S. investments, properties, and personnel in those areas of Latin America where North Americans were present and active.

A decisive step in the RF's initial involvement with Latin America began at two meetings of the executive committee of the IHC, held in July and October of 1915, when foundation officers approved the idea of undertaking a medical survey of the principal South American countries.[12] The main assumption that guided this decision was that the health of populations depended in the long run not on the RF's campaigns against tropical diseases but upon the quality of training of local medical practitioners.[13] After a series of inquiries, officers determined that Brazil would come under study first.[14]

The selection of Brazil, a country that attracted the attention of the RF for many years, arose from several factors. First, the United States and Brazil enjoyed very friendly diplomatic relations at the time.[15] Secondly, the foundation was deeply impressed by the outstanding work against yellow fever and bubonic plague of Oswaldo Cruz, the dynamic director of public health in Rio de Janeiro between 1903 and 1909.[16] After positive answers from the Department of State and the Brazilian government, the foundation created a survey commission that visited Brazil between January and May of 1916.[17] The commission examined fifteen Brazilian localities in terms of medical education, hospital organization, prevalent diseases, and public health agencies.[18]

Thanks to the prior experience of one of its members, the commission made extensive use of the dispensary method in the campaign against hookworm.[19] The dispensaries were mobile; staff set up a tent in each town and village they visited. There, they examined, treated and educated people, free of charge, in the belief that local practitioners would be capable of continuing the new work after an initial demonstration.[20] The essential staff consisted of a physician and lay assistants equipped with microscopes, drugs, charts, and other exhibit materials.

In a country such as Brazil—where the population was sparse, traveling was difficult, and large towns were scarce—the dispensary was an excellent means to make a favorable impression by bringing quick and temporary relief to the people. In their last camp in Belo Horizonte, the village priest and about five hundred people "dressed in the best they had" with "fireworks, flowers and the village band," greeted the American party.[21] However, after 1919, the tendency of the work of the IHC in Brazil was to leave the domain of demonstration against hookworm infection, and to take on more of the attributes of a state public health service.[22]

As a result of the 1916 commission and another visit by an RF officer in

1917, the foundation initiated a twofold program of work in Brazil: a hookworm campaign, and a cooperative program in the teaching of hygiene and pathology at the University of Sao Paulo. The RF selected the medical school at Sao Paulo rather than the medical school at Rio de Janeiro mainly because the officers were convinced that influential medical schools should be located in large, industrial cities that could provide resources for support, and Sao Paulo ranked as such a city. By the 1910s, Sao Paulo had become the most prosperous commercial center in Brazil. It specialized in production and marketing of coffee; thousands of Europeans had immigrated there, and it was developing an efficient and organized public health system.[23] In addition, the Sao Paulo medical school was new, under state rather than federal supervision, and unhampered by the many traditions of older schools, such as Bahia and Rio. The RF hoped that improvements could be introduced at Sao Paulo with less opposition than anywhere else, improvements that could influence medical education in Brazil and in the rest of Latin America.[24]

From 1916 until 1925, the RF assisted the Department of Pathology and the Institute of Hygiene of the University of Sao Paulo by selecting U.S. professors to send there, by paying supplementary salaries in connection with these professors, and by appropriating funds for the purchase of laboratory equipment. More important, negotiations began for placing the medical sciences on a full-time basis and to promote a new site for the school. (At the time, the school rented scattered buildings throughout the city.) In 1925, upon invitation from the RF, a commission of the Sao Paulo Faculty of Medicine visited medical institutions in the United States, Canada, and Europe, with a view to drawing up plans for a new physical plant for the school. When the Brazilian government approved the plans, the RF made an initial appropriation of U.S. $400,000 to be used for the construction of laboratories in the basic sciences.[25]

By 1916, the RF had committed itself to important undertakings in a number of foreign countries. Preliminary surveys preceding work on hookworm and other diseases had become standard practice. In May of 1916, Jerome Greene, the foundation's influential secretary of the early years, wrote a critical memorandum about these surveys. According to Greene, the surveys were too rapid, sketchy, and incomplete. He stipulated that future surveys should include a "thoroughly adequate knowledge of the whole social, educational and economic background of [a] country."[26] Greene's memorandum led to more-comprehensive surveys, of the kind first made in Argentina, Uruguay, and Chile between 1917 and 1919. These reports had an effect inside the RF because they were instrumental in the launching of the Division of Medical Education in 1919. After that year, the RF trustees assumed direct responsibility for the surveys, a decision that underlined the importance of the surveys as a tool of philanthropic planning.

Table 1. Countries Surveyed, by Year of Report

Year	Countries
1916	Brazil
1917	Argentina, Uruguay
1919	Brazil, Chile, Paraguay
1922	Guatemala, Nicaragua, El Salvador
1923	Colombia, Brazil, Mexico, Paraguay
1924	Brazil
1925	Argentina, Brazil, Paraguay
1926	Bolivia, Peru, Uruguay, Haiti
1927	Venezuela
1929	Cuba

Note: No reports were made in 1918, 1920, 1921, and 1928.
Source: See note 1.

No reports were made in 1918, because of World War I, but between 1919 and 1929, the RF surveyed at least once all the South American countries except Ecuador and the Guianas. Table 1 presents a chronology of the countries RF officials visited.

Six officers were responsible for the reports, and the three most active of the six were Richard Pearce, Robert Lambert, and Alan Gregg. These three were trained in first-class medical schools (where they acquired some persistent beliefs and attitudes about medical education), had served at well-equipped hospitals for at least a year, and eventually occupied very important positions inside the RF. Richard Pearce brought to the RF experience as an investigator and a teacher, and familiarity with administrative work. He spent a year studying at Leipzig, after receiving his M.D. at Harvard. Before working full-time for the RF, he served as director of the prestigious Bender Hygienic Laboratory at Albany and later as "professor of scientific medicine" at the University of Pennsylvania, where he played an important role in the professionalization of physiology.[27] By the early 1910s, Pearce was recognized as a leader in the promotion of a new research ethic in clinical medicine.[28]

Pearce visited six Latin American countries for the foundation. His reports were comprehensive and filled with detailed data on the countries' history, educational structure, and social texture. Pearce became one of the most-important officers of the foundation in 1919, when he was appointed director of the Division of Medical Education. In that capacity, he developed an international program concerned with the strategic funding of key medical schools around the world, foreign medical fellowships, and emergency aid to depressed, postwar Europe.[29]

Robert Lambert received his M.D. from Tulane in 1907 and pursued graduate study at Johns Hopkins and in Berlin. Before joining the foundation, he served as professor of pathology first at Columbia, and later at Yale, and gained hospital experience in New York hospitals. In 1922, the RF sent Lambert to El Salvador, responding to a request from the medical school there for a U.S. professor of pathology.[30] In Central America, he surveyed the conditions of a few medical schools, which were found to suffer from a severe scarcity of resources. He was later hired by the foundation to serve as professor of pathology in Sao Paulo from 1923 to 1925, and in Puerto Rico from 1926 to 1928. In 1929, he was appointed associate director for the medical sciences division of the RF.

During the period from 1922 to 1929, Lambert surveyed a total of eleven Latin American countries for the foundation. In one of his reports, he dealt with the question of how accurate an assessment an observer could make during a short visit—usually between one and ten days—for a survey. He defended a cursory five-day examination of Venezuela in the following terms: "Hastily acquired impressions may be faulty but they are more vivid and . . . often more accurate and just than the ideas of old residents."[31]

Alan Gregg surveyed only two Latin American countries—Mexico and Colombia—but expressed very articulately his opinions on Latin American culture and society. Trained at Harvard Medical School and Massachusetts General Hospital, Gregg served in France, treating war casualties, during World War I. Upon his return to civilian life in 1919, he was hired by the RF to work against hookworm in Brazil. (He recorded in a diary his impressions and anxieties while in that country.[32]) His career took a final turn in 1922, when Pearce offered him a job as his assistant in the RF Division of Medical Education. A few years later, Gregg became Pearce's associate director in charge of European operations,[33] and in the early 1950s, he became vice president of the RF.

The lack of precedents for international philanthropy partially explains why the foundation's officers tried to place Latin American realities into U.S. framework. Alan Gregg acknowledged the novelty of this experience in his Brazilian diary:

> I do not feel as if I had learned anything new for a long
> time—medically speaking. But on scenery, and cooking and customs
> and hookworm, and keeping accounts and learning to plan programs
> for other folks to follow—perhaps here is the difference. I certainly
> can't tell [if this] is progress because I haven't seen anybody to
> compare with. I feel like a fellah that comes out of a cloud up-side
> down and tries to find out where in the world the world is.[34]

Lambert, Gregg, and especially Pearce filed comprehensive reports on Latin America, usually following specific format. First, they provided information on the country's location, natural resources, form of government, and history. Second, they analyzed public health agencies and medical practice, presenting data on prevalent diseases, the ratio of doctors to population, the conditions of medical licensing, and special regulations concerning foreign physicians. Third, the reports described primary, secondary, and higher education in the country, including a brief history of medical schools and universities.

The bulk of each report concentrated on medical education, especially after 1922, when officers no longer strictly followed Greene's directions for comprehensive surveys. Description of a medical school typically included its academic and administrative organization, admission requirements, student fees, degree and course offerings, and physical plant, and a discussion of its importance as a medical center in its own country. Frequently, diagrams, plans, photographs, and appendices with copies of reprints also appeared in the report. Table 2 summarizes some of the information given by the reports produced between 1916 and 1929. RF officers usually gave special attention to three aspects of medical schools: laboratory facilities, full-time positions, and clinical facilities.

A View inside the Medical Schools

Although medical school laboratories appeared to be well supplied with apparatus and material in the most developed South American countries, RF officials were struck by the fact that clinicians occupied most of the chairs and facilities were not sufficient for the large number of students enrolled.[35] These conditions had come to exist in part because very large numbers of students enrolled during the first two years of medical education and because faculties included few full-time positions in the basic sciences. Frequently, the heads of laboratory departments were practicing physicians who confined their teaching to lectures and demonstrations, leaving any additional instruction to younger assistants. Initial enrollments were high because entrance requirements were low; administrators needed student fees to meet current budget expenses. The RF reports noted that medical schools in the United States enjoyed a very favorable faculty-student ratio, averaging one professor to three students, and that Latin American schools did not match that distribution.

In terms of laboratory instruction, the most thorough course work was offered in anatomy, bacteriology, and pathology. The two latter courses met for three hours every day. Other preclinical courses, such as histology and biological chemistry, met for only three hours a week or, in the case of physiology,

Table 2. Latin American Medical Schools According to RF Reports, 1917– 29

School	Year of Report	Faculty	Enrollment	Average No. Lab Facilities/ Class	Vols. in Library	Hospital Beds	Year of Founding
Buenos Aires	1917	140	3,173	50/325 bacteriology	48,000	700	1801
				80/800 physiology			
Cordoba	1917	34	434	8/73 physiology	n.a.	300	1878
				30/73 bacteriology			
Bahia	1924	31	580	22/100	19,360	n.a.	1808
Rio	1925	31	n.a.	12/900 anatomy	n.a.	n.a.	n.a.
Santiago	1919	32	948	50/348 anatomy	14,000	750	1860
				17/102 bacteriology			
Bogota	1923	30	398	22/170 anatomy	3,355	600	1867
				20/n.a. physiology			
Havana	1929	150	2,455	35/519 bacteriology	n.a.	1,300	n.a.
Mexico	1923	130	1,702	12/n.a. anatomy	10,000	625	1833
Lima	1926	40	487	30/110 anatomy	7,000	622	1856
				15/51 bacteriology			
				20/101 physiology			
Montevideo	1917	37	573	12/n.a. anatomy	11,000	500	1875
				16/n.a. physiology			

Source: See note 1.

four-and-a-half hours per week. Anatomy instruction consumed two years. During the first year, the student studied the human organism in general. The second-year course included the central nervous system, the sense organs, and the endocrine system. Only second-year students could do dissections, and laboratory work suffered from chronic difficulties in obtaining enough bodies for autopsies.

The reports rarely emphasized that, by rigid examinations, the medical schools reduced the number of students after the first two years, making for a high rate of attrition. Further, to avoid overcrowding in laboratory facilities, an innovative method of teaching was adopted. In Buenos Aires, in the pathology course, the students were required to study only forty preparations and could do so at any time they wished. The system permitted students to work individually, and ambitious students received the opportunity to study a larger amount of material. In addition, some first-class scientists, such as Bernardo Houssay in Buenos Aires, concentrated on the brightest medical students. Houssay selected his students from the large physiology classes. Taking into account the lack of a formal system for attracting talent to science, Houssay needed contact with "overcrowded" classes.

When the RF's official visited Houssay's institute, the Argentine scientist was becoming an international star because of his experimental work on the adrenal cortex and hypophysis. Houssay managed to group together the chairs of physiology, physical chemistry, and chemistry in what was called the Physiological Institute of the Medical School of the University of Buenos Aires. He served as chairman of the professors of these subjects, and devoted his time entirely to teaching and research, receiving the highest salary of any professor at the university.[36] Nevertheless, RF officials found Houssay's laboratory overcrowded, underequipped, and "far from clean."[37] After being escorted by Houssay through his laboratories, an RF officer wrote, "I got the impression that the disorder was probably chronic and that Houssay is one of those laboratory workers to whom housekeeping is a bore. The departmental library showed the lack of method evident elsewhere."[38]

Houssay's achievements were not common in Latin America. Typically, the Latin American professor of medicine was a practitioner first, and only incidentally a teacher or a researcher. Successful practitioners sought university positions because they conferred distinction, increased professional prestige, and eventually became steppingstones to political careers. In any case, a professor's stipend was usually too small to constitute a livelihood. All laboratories were either government-supported in connection with medical schools, boards of health, or hospitals, or were small, diagnostic laboratories for commercial gain. The medical schools existed as an integral part of the national educational machinery and depended upon the government for all expenses in excess of those covered by student fees and a small income from rentals of property. Private endowments were rare in the 1920s, and as a result, independent research laboratories were also rare.

However, some important features of the system in time made a difference. Younger scientists who were interested in research could remain relatively active in science by holding several positions in the university, thus gaining an income that permitted them to maintain little or no medical practice. This pattern typified an important compromise that fostered development of science in Latin America. Latin American scientists advanced their careers at a time when their societies were beginning to discover the importance of science in the national culture. Most researchers could not support themselves through only research work, scientific publications, and university teaching. In addition to working as researchers, they often had to take administrative positions or work as scientific journalists or inventors of scientific apparatuses. The persistence of multiple roles among scientists resulted from existence of a nonspecialized ethos that permeated the university, and other cultural institutions as well. RF officials suggested that multiple roles could only produce conflict and inefficiency. However, in order to work as scientists, Latin American physicians needed the visibility and support they found through their multiple university roles.

In many countries, the greatest need was the preservation of existing laboratories, personnel, and morale, rather than the creation or adaptation of new forms, especially in the face of an unpredictable future. In poor countries, science had to compete with other state agencies for scarce resources. Multiple roles and visibility became necessary. It was imperative for the pioneering Latin American researchers to keep scientific work as broad as possible. Their unorthodox approach to dealing with the uncertainty of a scientific career was difficult for the RF surveyors to understand.[39]

The promotion of full-time positions recalls the Flexner Report. The RF's surveys were deeply influenced by Abraham Flexner's 1910 report on U.S. medical schools, which was instrumental in the closing of most low-standard and poor medical schools, in the development of full-time positions and adequate laboratory facilities, and in promoting Johns Hopkins Medical School as the prototype of medical training.[40] Flexner was the chief proponent of the strict full-time system for clinical positions in the United States. Under this system, professors of clinical medicine would be paid fixed salaries, and they would not be permitted to earn income from the private practice of medicine. In actuality, U.S. universities adopted a more lenient attitude after 1920, allowing clinicians some private consultation.[41] The so-called geographical full-time usually meant that the school provided its professors with offices in an affiliated hospital where they conducted all their practice. No adaptation of the original Flexner scheme was tried in Latin America; there, the full-time concept was promoted in the medical schools and especially in the basic sciences.

RF surveyors found Latin American hospitals wanting in many respects, even in Buenos Aires, where the surveyors visited hospitals "excellently built and equipped and with adequate laboratories for routine diagnostic work."[42] RF officials raised one basic question: to what extent were the hospitals controlled by the medical schools? With the exception of the hospitals in Brazil, South American hospitals were administered by the Sociedades de Beneficencia, semiprivate charitable organizations that received subsidies from the government. Hospitals had no formal connection with medical schools. However, the hospital was used for university teaching, and the heads of the hospital services usually were professors in the medical schools.

Hospitals in Brazil were somewhat different in that they were administered by a lay brotherhood made up of a dues-paying membership of the Roman Catholic Church, which was engaged in a wide variety of charitable services.[43] The brotherhood existed in all the principal cities in Brazil, and the hospitals were called the Santa Casa de Misericordia (literally, Holy House of Mercy). The medical and administrative staff of the Santa Casa was appointed by the president of the brotherhood. As in the rest of South America, hospital staff members could serve as members of the faculty, and the most important services were used for university teaching.

In addition to the lack of formal control of the hospitals by the medical schools, foundation officers faulted Latin American hospitals' lack of good record keeping systems, female professional nurses, full-time interns and research laboratories, and "indeed everything we associated with modern hospitalization."[44] Male attendants or auxiliary nurses performed nursing services, usually under the supervision of French nuns. The hospitals used "interns" who were really medical students in their last years of attendance; these students served as clinical clerks, visiting at stated hours, taking histories, and performing physical examinations. These interns attended classes and did not live in the hospital except when on hospital duty—one day every two weeks or so.

However, some distinctive aspects of the Latin American system of hospital training deserve attention. Medical education in Latin America consisted of a prolonged course of seven years, three of which were passed practically in the hospital wards. One RF officer even entertained the idea that the system of three years of hospital training [in Argentina] might be preferable to the one-year American training of interns. He noted that young Argentine physicians had greater clinical experience than their American counterparts.[45]

The inside view of Latin American medical schools gained through the sur-

veys furnished the basis for the RF's goals for Latin American medical education: (1) limitation of the number of students; (2) establishment of full-time positions for the basic laboratory branches of anatomy, physiology, histology, bio-
chemistry, pathology, and bacteriology; and (3) establishment of direct faculty
control of the hospitals used for clinical instruction. In terms of individual
countries, the foundation concentrated its initial aid to South America in Brazil.[46]

It is interesting to note that the foundation chose Brazil rather than the
most advanced South American country of the early 1900s, Argentina. Some
probable reasons were that Argentineans had a centralized university structure
with only two medical schools, and were considered too proud and sensitive
to criticism. In contrast, Brazil had a federative form of government and state-
supported universities, characteristics that were seen as similar to those of the
United States. The selection of Brazil, and an examination of foundation records, suggests that RF officers sought to establish programs in Latin American
countries where fewer obstacles would exist to the actual use of new public
health measures and where officers anticipated a sound reception of the government bureaucracy and the medical elites.

RF officers found that medical schools in Chile, Argentina, and Uruguay
functioned reasonably effectively in clinical education. With respect to these
and other Latin America nations, the RF decided first to encourage a strong
public health program before undertaking any aid for full-time positions in
medical schools. At the same time, fellowships were established with an understanding between the RF and government officials that the people who received fellowships would, after returning to their home countries, be given
full-time positions.

After a major reorganization in 1928, the foundation expressed an interest
in encouraging investigative work in Latin America. This general interest in
funding individual scientists around the world continued to grow, and in January 1929, the Division of Medical Education became known as the Division of
Medical Sciences of the RF, foreshadowing an important change of policy.[47] As
a result, the foundation began to offer grants to Latin American researchers
working in the preclinical sciences, especially in physiology and neurophysiology.[48] The RF made another decision important to Latin America in 1928 when
it established a policy providing that aid for extremely backward countries
should be discontinued, because their "economic life cannot support modern
medicine."[49] Such a policy was consistent with the RF's evolving desire to
build elite institutions that would train leaders and set an example.

American perceptions of Latin American research emerged not only as a
result of the information gathered inside medical schools, laboratories, and
hospitals. Observations of a foreign culture and a different kind of society oc-

cupied an important place in the reports, and these observations contributed to RF assessments of Latin American science and medicine.

U.S. Perceptions of Latin American Society

Reflections on Latin American society and culture appeared in the reports initially as historical explanation of the causes of scientific and educational backwardness and reinforced the notion of the superiority of U.S. forms of scientific and social organization. The reports frequently criticized the heavy dependence of Latin American universities on civil authority, a dependence that began in the nineteenth century when the state took over control of the old universities of colonial and religious origin. The reports also complained about the strong influence among Latin American universities of French educational and cultural patterns. The French influence, in contrast to the German example, was thought to result in emphasis on verbal expression, part-time and didactic teaching, and little regard for experimental work. As an example of the persistence of that influence, an RF officer noted that of the 13,715 works read during a year in the library of the medical school of Buenos Aires, 52 percent were in French, 2 percent in German, and not even 1 percent in English.[50]

The openly racist overtones of the historical reflections were characteristic of scientific and social thinking in the United States, Europe, and Latin America during the 1920s, due in part to the influence of eugenics ideas.[51] In the conclusions of the report on Colombia, Alan Gregg speculated on the "clues to the South American mind." The first clue was geographical diffusion. According to Gregg, social solidarity did not exist among Latin Americans because the Spanish conquerors were marauders, not settlers accompanied by families. In addition, social solidarity was never established because Spanish central authority tried to cover a territory that was too vast, and thus the repression of individualism dominated government action.[52] Gregg lamented fast and vast settlement, praising instead the slower pattern of colonization by which Europeans propagated their "social order and sense of responsibility for each other." Of course, he was mirroring the conventional wisdom on the history of the United States.[53]

The second clue to the South American mind was "racial dilution," the "extensive biological dilution of the European race." This idea was not original with Gregg, and held that the Europeans who had conquered South America had diluted their abilities by mixing their bloodlines with those of the colored, indigenous peoples. According to Gregg, the fact that the Latin American Indians were not exterminated, but "enslaved, in reality or in disguise," created the conditions by which the Europeans were absorbed *"by the race they conquered"* (Gregg's emphasis). "Many regions," he continued, "have been populated by a mulatto and mestizo group which cannot be compared with the Latin side of their ancestry in ability and traditions." For example, he found Mexicans "phlegmatically submissive . . . of low vitality and poor stock." Moreover, the Mexican Indian was "a secondary race like the Javanese in the Dutch East Indies," one with a "natural lack of interest in education."[54] For Gregg, the lack of a "uniform race" went a long way toward explaining the underdevelopment of science, medicine, and education in Latin América in the 1920s.

The RF's reports used isolated elements of Latin American culture, society, and education as indicators of modernness or backwardness, paying little attention to local meaning and function. The assumptions in these reports were based on an elitist reading of the history of science and society in the United States, and RF officers considered the assumptions valid for planning programs for all societies.

Discussion

RF officials proceeded from a main assumption—that the most-advanced Latin American institutions and societies could replicate the development of their counterparts in the United States. There were references to Sao Paulo as a "California town," to the Argentinean attitude toward medicine as "not greatly different from that in the U.S. thirty to forty years ago," and to Brazil as "our own country of 1830 to 1850 all over again."[55] As a result, scientific and social development was seen as a linear path along which all countries traveled. A second principal assumption held that change was produced by external stimulus. There were constant references to the role that German-trained U.S. students played in the modernization of medical education and research in the United States. Foundation officers became sincerely convinced that the RF would play the same role in Latin American science and medicine that Germany had played in the reorganization of U.S. science.[56]

With the advantage of having today a better historical understanding of the differences between the U.S. and Latin America, and a sense of the development of science in the periphery, we can analyze these assumptions and the different functions of the early surveys and RF reports of the Progressive Era. First, there was the problem of overstating the German influence in the United States. U.S. medical schools borrowed selectively from the model of the German university. One of the main departures from the German model was the department system, which tended to disperse power among several full-time professors. Professors in German institutes enjoyed more power than their American counterparts.[57] As was the case with the full-time system, no Latin American adaptation was considered.

Second, there was the problem of confidentiality. Organizers of surveys in the United States during the Progressive Era had a clear idea of the public role their work would play. These surveys combined social research and popular education in an attempt to bring about political change. The findings of surveys in that period were widely publicized in books, newspapers, and magazines.[58] The Flexner Report was written in close collaboration with the predominant medical professional organization of the time and reflected the current needs of the medical elite for higher standards. The Carnegie Foundation printed 15,000 copies of the Flexner Report for distribution at no charge.[59]

In contrast, the RF reports on Latin American medicine and science were delivered to the New York office of the foundation and circulated only among the officers, staff, and trustees. Confidentiality was defended with the argument that these reports contained information given in confidence, the release of which might embarrass others as well as the foundation. However, secrecy eventually undermined RF attempts to encourage medical reform, and limited public awareness of the foundation's general program. This attitude can be understood in light of the RF's underlying belief that change should occur from above.

The criteria that foundation officials used in incorporating judgments into the reports on Latin American schools were influenced by Flexner's reflections on what he saw as substandard U.S. schools. His report held that the United States suffered from an overproduction of badly trained medical personnel because of the large number of inadequate medical schools. Flexner advocated producing fewer and better physicians by reducing drastically the number of U.S. medical schools and by letting the better-equipped and more efficient schools survive.

As Mary Brown Bullock has argued with respect to China, the Flexner Report represented the culmination of an elitist process in U.S. higher education, one that involved the creation of a number of elite and very influential institutions, and a transition from a multiplicity of standards to uniform standards.[60] This process was not necessarily occurring in other regions of the world. The change occurred in the United States not only because of university regulations but because of a general recognition of the importance of experimental science, the availability of private endowments, and the acceleration of medical professionalization. Latin American universities were not living in such an environment in the 1920s; they were experiencing a populist period, which led to the so-called movements of university reform.[61] In several institutions, students and alumni secured control of one-third of the representatives in the legislative bodies of the universities, and acquired the right to pass upon the merits of new faculty members before they could be appointed by the school. The students pushed to keep enrollments high, and fought "elitist" efforts to raise laboratory fees and tuition.

The RF reports also assumed that creating examples of excellence in Latin America would initiate a process of imitation and thus improvement in the standards of medical schools. There was little comprehension that the medical education system was distinct in Latin America, especially in terms of the system of proprietary schools that prevailed in the United States during the nineteenth century (and which Flexner criticized in his report). As Daniel Levy has pointed out, universities in Latin America operated under strict government control, so that significant standardization occurred, while each medical school in the Unites States was substantially responsible for itself and competing with the others within a highly stratified system. In contrast, among Latin American universities, the drive to compete was low because virtually all policy decisions were made nationally for the whole system of higher education.[62]

The foundation's lack of understanding of the relationship between the Latin American states and medical schools reveals that, in spite of the comprehensive reports on society and government, the RF was generally too timid in asking social questions. The surveyors made few inquiries about the role played by the ruling elites in the social system, the external dependence of these countries, and the exploitation of the Indians. Instead they tried to find ways to induce institutional and individual change with little reference to the sociocultural matrix.[63] This disregard of the social situation was dramatic in the foundation's assessment of the kind of medicine and public health Latin America required of the 1920s.

The foundation insisted that Latin American countries needed a small number of doctors highly trained in the basic sciences and based in urban hospitals attached to medical schools. Eventually, this policy encouraged the formation of a conservative medical community and the development of an expensive medical system that was inaccessible to the majority of the people.[64] The RF, and the medical elites of the Latin American countries, gave low priority to the creation of a body of sanitary inspectors and assistants and rejected cooperation between trained physicians and the native healers. In some countries, such as Mexico, Guatemala, Brazil, and Peru, because of the persistence of Indian and black cultures, traditional healing practices had continued with little modification or were combined with modern drugs, and there was no strong movement toward resolving what RF officials viewed as a logical inconsistency. In consequence, a pattern of medical professionalization was promoted that implied the displacement of native practitioners.

In many Latin American countries, where the majority of the population lived in rural areas, trained physicians concentrated in the urban centers. There was little indication of an oversupply of doctors (as Flexner argued for the United States) but rather a general shortage of physicians and other medical resources throughout Latin America. For example, in Guatemala in the 1920s, a population of 2,500,000 inhabitants had only 132 doctors, 74 of whom worked in Guatemala City.[65] Another example of the difficulty of not using native resources was the obstacle encountered by the foundation's hookworm campaign in Brazil, which could not be carried to a self-perpetuating stage because of the lack of sufficient native physicians.

Refusing to deal with native healers led to problems of cultural resistance, as described by Gregg in his Brazilian diary:

> We were testing how people's blood was by pricking their ears and seeing how red the drop of blood was on a piece of blotting paper. The curandeiro or medicine man there did not like to see his patients treated by anybody else so he told them that we were selling the blood on the bloodletting paper to the Devil. So all the people ran home and we had nobody left to work with in that town.[66]

The surveys suggest a contrast between the philanthropic goals the RF derived from the U.S. experience, and the needs of societies with specific social and cultural traditions. This contrast was difficult to perceive and overcome in the 1920s due to many factors, including the remarkable flexibility of field officers who adjusted the policies of the foundation, and to the obstacles created by some Latin American medical elites who were more eager than the Americans to imitate U.S. institutions in their own countries. Eventually, time and experience were instrumental in revealing the contrast for some officers.

In the late 1940s, Robert S. Morrison, the RF's assistant director of medical science, evaluated the early assumptions of the foundation in the third world. First, he felt, making the "peaks higher" was a motto that overlooked local conditions; and second, it was difficult to see how the promotion of first-class doctors, many times badly trained, would be more functional to less-developed societies than "first class second class doctors." He also wrote,

> None of us who have been trained in Western medicine have given sufficient thought to the problem of adapting it for use in those large areas of the world which cannot possibly support the expense of medical practice as we know it. The idea of any so-called "second class" type of medicine is so repugnant to our mores that everyone seems to avoid thinking about the obvious fact that it requires a very high level of economic advancement to support a first class medical service in our sense of the word.[67]

The vision that emerges from the reports suggests a development framework in which everything Western appears to be modern, harmonic, rational, and probably superior, whereas everything non-Western was presented as traditional, incoherent, outdated, and presumably inferior. This dichotomous framework dominated the design of RF programs in Latin America during the 1920s and 1930s. Social scientists try today to answer a fundamental question: to what extent are all the elements of traditional institutions inconsistent with the process of change and modernization?[68] The RF reports on Latin America suggest that new scientific ideas and practices might be better consolidated in the third world not by the dislocation of all indigenous values but rather by allowing some reinterpretation as well as some coexistence with dissonant cultural elements.

Notes

1. These countries were Argentina, Bolivia, Brazil, Chile, Colombia, Cuba, El Salvador, Guatemala, Haiti, Mexico, Nicaragua, Paraguay, Peru, Uruguay, and Venezuela. The reports are kept in the Rockefeller Archive Center (hereafter RAC) in Record Group 1.1. of the collection Rockefeller Foundation Archives (hereafter RFA). They are: "Medical Education in Argentina by Dr. Richard M. Pearce, 1917, Dr. R. A. Lambert, 1925," Series 301, Box 2, Folder 18; "Medical Education in Bolivia by R. A. Lambert," Series 303, Box 1, Folder 9; "Medical Education in Brazil, Diaries, Memoranda, Notes, Reports by the Medical Commission to Brazil, 1916, Lewis W. Hackett, 1919, Dr. Robert A. Lambert, 1923–25, George K. Strode, 1925–26," Series 305, Box 2, Folder 15; "Medical Education, Medical Relief and Public Health in Chile by Dr. Richard M. Pearce, 1919," Series 309, Box 2, Folder 22; "Medical Education in Colombia by Dr. Alan Gregg, 1923," Series 311, Box 2, Folder 19; "Notes on the Faculty of Medicine and Natural Sciences of the National University, Colombia. By Dr. W. M. Monroe, 1923," Series 311, Box 3, Folder 19; "Medical Education in Cuba, Dr. R. A. Lambert, 1929, 1946," Series 315, Box 1, Folder 2; "Medical Education in Cuba, by the Information Service, 1927–1928 (Compiled from Recent Publications)," Series 315, Box 1, Folder 1; Robert A. Lambert, "Medical Education in Guatemala: Report on the Guatemala Medical School, 1922," Series 319, Box 1, Folder 1; Richard M. Pearce, "Medical Education in Haiti, 1926," Series 320, Box 1, Folder 3; "Medical Education in Mexico by Alan Gregg—1923, Dr. R. A. Lambert—1936–1941," Series 323, Box 13, Folder 95; Robert A. Lambert, "Medical Education in Nicaragua: Reports on the Schools at León and Granada, 1922," Series 325, Box 1, Folder 1; "Medical Education and Public Health in Paraguay by Drs. Pearce, Lambert, Hackett and Strode, 1919–1925," Series 329, Box 1, Folder 3; "Medical Education in Peru by R. A. Lambert, 1926," Series 331, Box 3, Folder 19; Robert A. Lambert, "Report on the Salvadorian Medical Schools, 1922," Series 333, Box 1, Folder 3; "Medical Education in Uruguay, 1917–1925," Series 337, Box 1, Folder 1; "Medical Education in Venezuela, by R. A. Lambert, 1927," Series 339, Box 1, Folder 4.

2. As Smith points out, 2,700 such surveys were undertaken in the United States from 1900 to 1928, on subjects ranging from education and sanitation to crime. See James A. Smith, *The Idea Brokers, Think Tanks and the Rise of the New Policy Elite* (New York: Free Press, 1991), 41.

3. This work is an addition to Thomas Glick's interpretation, presented in "The Spanish Laboratory Crisis of the 1920s: Rockefeller Foundation Officers Assess the Culture of Scarcity," paper presented to the History of Science Society Meeting, Seattle, Washington, October 1990. According to Glick, when the RF's experts assessed Spanish scientists, they did so according to U.S. criteria that screened out local conditioning factors.

4. The name of the International Health Commission was changed to the International Health Board in 1916, and again in 1927 to the International Health Division. For a history of this division, see Lewis Hackett, "Once upon a Time," *American Journal of Tropical Medicine and Hygiene* 9, 2 (1960): 105–15, and Raymond B. Fosdick, *The Story of the Rockefeller Foundation* (New Brunswick, N.J.: Transaction Publishers, 1989), 58–70.

5. These colonies included Egypt, Ceylon, and the Malay states. Rockefeller Foundation, International Health Board, *Itinerary of Director General through Egypt, Ceylon, Federated Malay States and Philippine Islands*, March 8, 1914 to July 7, 1914, Rockefeller Archive Library, International Health Board 2, RAC. The selection of areas under British authority reflects the close relationship that the RF had with England from its beginning.

6. Work was undertaken successively in British Guiana, Trinidad, Grenada, Saint Vincent, Saint Lucia, and Dutch Guiana. International Health Commission, "Committee to Study and Report on Medical Conditions and Progress in Brazil," in "Medical Education in Brazil. Diaries, Memoranda, Notes, Reports by the Medical Commission to Brazil, 1916, Lewis W. Hackett, 1919, Robert A. Lambert, 1923–25, George K. Strode, 1925–26," RFA, 1.1, Series 305, Box 2, Folder 15, RAC.

7. Again the foundation began its campaign on a small scale and in a "quiet way." Work was undertaken in Panama, Costa Rica, Guatemala, and Nicaragua. "Medical Education in Brazil."

8. When the RF initiated its work, there was no other effective international public health agency working in Latin America. The International Sanitary Bureau, created in 1904 by various states of the Americas, was only a facade shielding a very weak bureaucracy. George Vincent, Diary, March 6, 1923, RFA, R.G. 12.1, Box 65, RAC.

9. Wickliffe Rose, "Memorandum on Yellow Fever: Feasibility of Its Eradication," October 7, 1914, RFA, R.G. 5, Series 300, Box 22, Folder 135, RAC.

10. Recent studies on these campaigns are Armando Solorzano, "The Rockefeller Foundation in Mexico: Nationalism, Public Health, and Yellow Fever (1911–1924)" (Ph.D. dissertation, University of Wisconsin, Madison, 1990); and Marcos Cueto, "Sanitation from Above: Yellow Fever and Foreign Intervention in Perú, 1919–1922," *Hispanic American Historical Review* 72 (1992): 1–22.

11. Discussions of the changes in Latin America during the early twentieth century appear in Leslie Bethell, *Latin America, Economy and Society, 1870–1930* (Cambridge: Cambridge University Press, 1989); and Thomas E. Skidmore and Peter H. Smith, *Modern Latin America* (New York: Oxford University Press, 1984).

12. International Health Commission Minutes, Brazil Medical Commission, July 1, 1915, in "Medical Education in Brazil."

13. "Medical Science, Program and Policy," extract from agenda for special trustees meeting, April 11, 1933, RFA, R.G. 3, Series 906, Box 1, Folder 8, RAC.

14. This survey was preceded by surveys on medical and educational activities in China; see *Report of the China Medical Commission to the Rockefeller Foundation* (New York: Rockefeller Foundation, 1914). China Medical Board, R.G. 2, Box 11a, Folder 89a, RAC.

15. U.S.-Brazilian relations thus stood in contrast to relations between the United States and revolutionary Mexico, where no large program was developed until the early 1940s, after the decline of the intense Mexican nationalism that characterized the period following the 1910 revolution.

16. In 1909 the Brazilian government transformed Cruz's laboratory into a full-fledged government institute, where microbiologists and protozoologists carried out scientific experi-

ments that achieved international recognition. See Nancy Stepan, *Beginnings of Brazilian Science: Oswaldo Cruz, Medical Research and Policy, 1890–1920* (New York: Science History Publications, 1981).

17. The commission was formed by Richard M. Pearce, Bailey K. Ashford, and John A. Ferrell. Wickliffe Rose to William Phillips, October 6, 1915, RFA, R.G. 1.1, Series 305, Box 2, Folder 15, RAC.

18. "Medical Education in Brazil." Specific reports were written for only eight localities: Bahia, Curytiba, Rio de Janeiro, Porto Alegre, Rio Grande do Sul, Pelotas, Florianopolis, and Juiz de Fora.

19. The commissioner with experience was Bailey K. Ashford, who had used the dispensary method in Puerto Rico in 1904. He was the discoverer of *Necator americanus*, the internal parasite that causes hookworm disease in the New World. Ashford's work in Puerto Rico was brought to the attention of John D. Rockefeller, who then created the Rockefeller Sanitary Commission. See Annete B. Ramirez de Arellano, "The Politics of Public Health in Puerto Rico, 1926–1940," *Revista de Salud Pública de Puerto Rico* 3 (1981): 35–73.

20. The dispensary became very popular in the United States during the nineteenth century as an alternative to the hospital in providing medical care for the urban poor. See Charles Rosenberg, "Social Class and Medicine in Nineteenth Century America: The Rise and Fall of the Dispensary System," *Journal of the History of Medicine and Allied Sciences* 24 (1974): 32–54; and Charles Rosenberg, ed. and intro., *Care for the Working Man: The Rise and Fall of the Dispensary, An Anthology of Sources* (New York: Garland, 1989).

21. Bailey K. Ashford, "Narrative of the Medical Expedition of the International Health Commission to Brazil," in "Medical Education in Brazil,"21.

22. Lewis Hackett, "Relief and Control of Hookworm Disease in Brazil, from November 22, 1916, to December 31, 1919," RFA, R.G. 5, Series 2, Box 24, Folder 148, RAC. This was partially due to the decline of the dispensary system in the United States; Rosenberg, "Social Class."

23. The emergence of the public health system in Sao Paulo is analyzed in Luiz Antonio de Castro Santos, "Power, Ideology and Public Health in Brazil, 1889–1930" (Ph.D. dissertation: Harvard University, 1987).

24. International Health Commission, Minutes, in Sao Paulo Faculty of Medicine, "Historical Record," vol. 1, RFA, R.G. 1.1, Series 305, Box 3, Folder 22, RAC.

25. The government agreed to furnish the remaining buildings necessary for medical instruction, including a teaching hospital. By 1930 the foundation's contributions to the Sao Paulo medical school amounted to U.S. $1,000,000. Fasciolo to Miller, June 19, 1957, RFA, R.G. 1.2, Series 301, Box 5, Folder 34, RAC.

26. The memorandum was presented at a trustees meeting, Jerome D. Greene, "Surveys of Foreign Countries," May 24, 1916, RFA, R.G. 3, Series 300, Box 21, Folder 164, RAC.

27. George Vincent, David L. Edsall, Howard T. Karsner, et al., *Richard Mills Pearce Jr. 1874–1930* (New York: Rockefeller Institute of Medical Research, 1930).

28. One of Pearce's essays was a historical reflection on the place of research in medical education from antiquity to the twentieth century. His summary of the development of American physiology during the late nineteenth century presented four stages: first, inquiry by individual scientists; second, application of exact methods of chemistry, physics, and biology to medicine; third, the development of a community of scientists; and, fourth, the medical application of the knowledge gained by research. Richard M. Pearce, "Research in Medicine," in Richard M. Pearce et al., *Medical Research and Education* (New York: Science Press, 1913), 1–67. See also W. Bruce Fye, *The Development of American Physiology; Scientific Medicine in the Nineteenth Century* (Baltimore: Johns Hopkins University Press, 1987), 222–25.

29. Alan Gregg, "The Work of the Rockefeller Foundation in Medical Education and the Medical Sciences, 1920 to 1929, under the Direction of Richard Mills Pearce Jr., M. D.," *Rockefeller Foundation Quarterly Bulletin* 5 (1931): 358–75.

30. During Lambert's stay in El Salvador, the dean of the medical school made an appeal for U.S. $100,000, to reconstruct the buildings damaged by an earthquake. However, no appropriation was made. Lambert to Gregg, November 17, 1922, RFA, R.G. 1.1, Series 333, Box 1, Folder 2, RAC.

31. Robert A. Lambert, "Medical Education in Venezuela, by Robert A. Lambert, 1927," RFA, R.G. 1.1, Series 339, Box 1, Folder 4, RAC. There was some precedent for the rapid work of the surveyors. Flexner also made a rapid visit to each medical school for his 1910 report.

32. Alan Gregg, Brazil Diary, 1919, Alan Gregg Papers, Ms C 190, Box 5, History of Medicine Division, National Library of Medicine, Bethesda, Md.

33. Theodore M. Brown, "Alan Gregg and the Rockefeller Foundation's Support of Franz Alexander's Psychosomatic Research," *Bulletin of the History of Medicine* 61, 2 (1987): 155–82.

34. Gregg, Brazil Diary, 1919.

35. For example, only a fraction of the students in the medical school at Bahia, Brazil, could work in the laboratory at one time. "Medical Education in Brazil."

36. He received 1,200 Argentinean pesos, about U.S. $500 per month. "Medical Education in Argentina by Richard M. Pearce, 1917, Dr. R. A. Lambert, 1925," RFA, R.G. 1.1, Series 301, Box 2, Folder 18, RAC.

37. Vincent, Diary, 1923, May 11, 1923, RFA, R.G. 12.1, Box 65, RAC.

38. "Medical Education in Argentina." This impression might have affected the fact that only after 1936 did Houssay receive significant support from the foundation; see Marcos Cueto, "The Rockefeller Foundation's Medical Policy and Scientific Research in Latin America: The Case of Physiology," *Social Studies of Science* 20, 2 (1990): 232.

39. According to one RF officer, "no one [in Buenos Aires] appears to see the value of the full-time laboratory." "Medical Education in Argentina."

40. Abraham Flexner, "Medical Education in the United States and Canada," *Bulletin No. 4* (New York: Carnegie Foundation for the Advancement of Teaching, 1910). Works on Flexner include: Steven C. Wheatley, *The Politics of Philanthropy, Abraham Flexner and Medical Education* (Madison: The University of Wisconsin Press, 1988); Howard S. Berliner, *A System of Scientific Medicine: Philanthropic Foundations in the Flexner Era* (New York and London: Tavistock Publications, 1985); Robert P. Hudson, "Abraham Flexner in Perspective: American Medical Education 1865–1910," *Bulletin of the History of Medicine* 56 (1972): 545–61; and Thomas Neville Bonner, "Abraham Flexner and the Historians," *The Journal of the History of Medicine and Allied Sciences* 45 (1990): 3–10.

41. The University of Cincinnati was one of the early beneficiaries of a new RF policy which came to be known as "geographical full-time." See Ellen Corwin Cangi, "Abraham Flexner's Philanthropy: The Full-time System in the Department of Surgery at the University of Cincinnati College of Medicine, 1910–1930," *Bulletin of the History of Medicine* 56 (1982): 160–74.

42. "Medical Education in Argentina."

43. For the origins of this system, see A. J. R. Russell-Wood, *Fidalgos and Philanthropists, the Santa Casa da Misericordia of Bahia 1550–1755* (Berkeley and Los Angeles: University of California Press, 1968).

44. "Belo Horizonte," in "Medical Education in Brazil."

45. "Medical Education in Argentina."

46. In the early 1920s, Pearce wrote a letter to the president of the Rockefeller Foundation, explaining that there was no urgent need of special institutional assistance in Brazil other than the Sao Paulo Medical School in Brazil. Pearce to Vincent, May 4, 1922, Pearce Diary, Vol. 2, RFA, R.G. 12.1, Box 5, RAC.

47. The changes that the RF experienced during the late 1920s are reviewed in Robert E. Kohler, *Partners in Science: Foundations and Natural Scientists 1900–1945.* (Chicago: University of Chicago Press, 1991), 233–62.

48. "Medical Sciences, Program and Policy, History and Future Program," extract from agenda for special trustees meeting, April 11, 1933, RFA, R.G. 3, Series 906, Box 1, Folder 8, RAC.

49. Report of the Special Committee on the Division of Medical Education of the Rockefeller Foundation, November 9, 1928, RFA, R.G. 3, Series 906, Box 1, Folder 7, RAC.

50. "Medical Education in Argentina." The surveyor complained that all medical schools were organized and conducted after the French model.

51. See J. Edward Chamberlain and Sander L. Gilman, eds., *Degeneration: The Dark Side of Progress* (New York: Columbia University Press, 1985); and Nancy Leys Stepan, *The Hour of Eugenics: Race, Gender and Nation in Latin America* (Ithaca: Cornell University Press, 1992).

52. "Medical Education in Colombia by Alan Gregg, 1923," RFA, R.G. 1.1, Series 311, Box 2, Folder 19, RAC, 1–5.

53. A third clue advanced by Gregg lay in the Catholic church, which in Latin America controlled many institutions and charities. According to Gregg, the traditional custom of giving large government and private donations to the church constituted the principal obstacle for the emergence of private endowments. Ibid.

54. "Medical Education in Mexico, by Alan Gregg, 1923," 11, RFA, R.G. 1.1, Series 323, Box 13, Folder 95, RAC.

55. Gregg, Brazil Diary, 1919.

56. The role played by German universities in training U.S. students of medicine was studied by Thomas Neville Bonner, *American Doctors and German Universities, A Chapter in Intellectual Relations, 1870–1914* (Lincoln: University of Nebraska Press, 1963).

57. Two works that emphasize the adaptation of German models to American universities are: John Higham, "The Matrix of Specialization," and Fritz Ringer, "The German Academic Community," in Alexandra Oleson and John Voss, eds., *The Organization of Knowledge in Modern America, 1860–1920* (Baltimore and London: Johns Hopkins University Press, 1979), 3–18 and 409–29.

58. Smith, *The Idea Brokers*, 41.

59. Berliner, *A System of Scientific Medicine*, 118.

60. Mary Brown Bullock, *An American Transplant: The Rockefeller Foundation and Peking Union Medical College* (Berkeley: University of California Press, 1980).

61. These movements began in 1918 with a student strike in Cordoba, Argentina, and in the following years, the movement extended to much of Latin America.

62. See Daniel Levy, "Centralizing the University," in *To Export Progress: U.S. Assistance to Latin American Universities*, August 1990, draft, 2.

63. There were only general references such as, "there is much in Mexico that is reminiscent of the feudal system." "Medical Education in Mexico."

64. An illuminating study of this problem for another region of the third world is Peter J. Donaldson, "Foreign Intervention in Medical Education: A Case Study of the Rockefeller Foundation's Involvement in a Thai Medical School," *International Journal of Health Services* 6, 2 (1976): 251–70.

65. Robert A. Lambert, "Medical Education in Guatemala: Report on the Guatemala Medical School, 1922," RFA, R.G. 1.1, Series 319, Box 1, Folder 1, RAC.

66. Gregg, Brazil Diary.

67. Morrison to Gregg, November 23, 1949, RFA, R.G. 3, Series 906, Box 1, Folder 5, RAC.

68. At least one important case of articulation between Indian and Western Medicine has been studied; see Marcos Cueto, "*Indigenismo* and Rural Medicine in Peru: The Indian Sanitary Brigade and Manuel Nuñez Butrón," *Bulletin of the History of Medicine* 65 (1991): 22–41.

2

Nationalism and Public Health

The Convergence of Rockefeller Foundation Technique and Brazilian Federal Authority during the Time of Yellow Fever, 1925–1930

Steven C. Williams

THE REINTRODUCTION OF yellow fever to Brazil's coastal urban centers during the 1920s raised a number of issues of national concern. The Rockefeller Foundation's direct participation in Brazil's public health affairs placed the organization squarely in the middle of a brewing national crisis. At odds with Brazilian methods of combatting the disease, the foundation sought to impose its own approach, which was designed to coordinate an otherwise fragmented Brazilian yellow fever policy. The most difficult hurdle involved wrestling responsibility for the disease from individual Brazilian states that derived a good measure of political benefit from attempts to control yellow fever. The Rockefeller Foundation's antilarval approach, in contrast to the traditional antimosquito methods of the states, not only proved more successful in eliminating yellow fever from certain cities where infections became especially troublesome, but it also aided the federal government's ambitions to take over more and more of the health responsibilities formerly ceded to individual Brazilian states. As this chapter shows, however, the implementation of foundation methods was fraught with hazards—both medically and politically significant—which exposed a number of deficiencies inherent in the foundation's approach to yellow-fever control.

Public health authorities in Brazil had for decades confronted the deadly specter of yellow fever with few effective measures to combat it. Since its introduction into the country in 1849, official response usually remained locally confined, rarely extending beyond the limits of an affected urban center. The central government in Rio de Janeiro had legal obligations to prevent the introduction of foreign diseases into the various port cities of the country, but in practice such responsibilities were exercised only after the introduction of infection. In such an environment, epidemic diseases like yellow fever quickly became facts of life in late-nineteenth-century Brazil.

The linkages between parasitology, entomology, and epidemiology supported the development of Carlos Finlay's active-intermediate-host concept. The Cuban physician posited a mosquito, later known as *Aedes aegypti*, was the transmitting agent that delivered the disease from human to human.[1] Though Finlay proposed the theory during the 1880s, field testing did not occur until the turn of the century when the North Americans employed antimosquito measures in Havana. By eliminating *Aedes aegypti* adults, the microbial agent causing the disease was deprived the opportunity to infect susceptible human hosts. One city after another throughout the Americas successfully implemented Finlay's principles, thus dramatically reducing, if not completely eliminating, the immediate threat of yellow fever.

Brazilian cities were no exception to this trend. Both Santos and Rio de Janeiro effectively eliminated yellow fever by imposing rigid, door-to-door fumigation techniques designed to kill adult *Aedes aegypti*. Even with the successful eradication of yellow fever in these strategically prominent southern port cities, however, Brazilian public health authorities feared continued contact with infected northern ports might eventually reintroduce the disease to the southern states of the country. Attempts to rectify the situation included an effort on the part of Oswaldo Cruz—already a hero for his work in Rio de Janeiro—to eradicate focal points of the disease in particular cities of the north, like Belém in the state of Pará.[2]

Cruz's methods produced immediate and dramatic results. By using an elaborate grid system, an affected city was divided into zones, each of which had teams of *mata-mosquitos* (literally, "mosquito killers") that responded to instances of diagnosed yellow-fever attack. Squads using imported Clayton pumps fumigated all buildings within the vicinity of an outbreak. Because of *Aedes aegypti's* domestic characteristics, the effect was so thorough as to cause a disruption in the crucial transference of the virus between hosts—man and mosquito. Given wide enough dispersions, it was theoretically possible to break the viral chain in an affected area. Public health officials like Oswaldo Cruz were well informed of specific zones of infestation and gave particularly intense scrutiny to these *focos* of disease. Although such methods broke the chain of infection, they never entirely destroyed mosquito populations in an affected urban center. *Aedes aegypti's* range was often dispersed throughout the sprawl of a city: as urbanization expanded, so, too, did the reach of the mosquito. Once the crisis of yellow-fever infection retreated and officials relaxed their vigilance, it was not uncommon for *Aedes aegypti*—this time uninfected with the yellow-fever virus—to come back to areas where previous populations of the mosquito had been wiped out by fumigation. Reinfestation, in fact, was often so intense that, as a service to irritated residents, public health officials at the urging of local politicians occasionally deployed the Clayton fumigant pumps to reduce annoying mosquito populations. Local public health authorities often underscored the value of reducing mosquito populations to local political leaders as a means to quell potentially rebellious residents who—fed up with an overall lack of public services—might, otherwise, become disruptive.[3]

The consequences of fumigation were threefold. First, fumigation killed mosquitoes in massive numbers. Second, it resulted in political support for those local city bosses able to offer urban constituents temporary relief from the irritations of mosquitoes, infected or not. Third, it legitimized state medicine by demonstrating that, given proper impetus, public health officials could produce dramatic results despite the staggering disease realities of Brazilian cities.[4]

Fumigation, though, was costly, imperiling the outright institutionalization of the *mata-mosquito* squad in Brazilian public health practice. Geography further complicated matters, especially in the north of the country. Unlike the south and southeast, which enjoyed relatively cooler winters, northern regions produced ambient temperature conditions capable of supporting *Aedes aegypti* year-round. However, the temporary nature of antimosquito fumigation could not permanently eradicate *Aedes aegypti*.

The limitations of a federalist political system, which placed the burden of responsibilities for public health in the hands of individual state governments, contributed much to the ad hoc nature of official response. Though federal interest in poorer regions like the northeast increased especially during the 1920s,[5] the national government in Rio became involved in the affairs of individual states only on a limited basis. In fact, until the 1920s, the federalist political arrangements of the First Republic greatly impeded the implementation of nationally integrated public health programs on a consistent basis.

Severe budget cutbacks caused by the outbreak of the First World War in 1914 dismantled or undermined nearly all remaining antimosquito programs in Brazil.[6] The war in Europe greatly curtailed the abilities of federal and state governments to collect duties to pay for such programs. In spite of the situation, most concerned officials took comfort in the low incidence of infection. However, not everybody expressed equal optimism. Calling the extinction of yellow fever a "dominating necessity," Brazil's minister of the interior in 1915 demonstrated extreme concern about northern focal points of infection, warning that continued infestation posed "a perennial source of incalculable evil."[7] The possibility of a chance introduction of the disease into southern Brazil greatly preoccupied a number of informed observers. Over a half-century of experience had taught Brazilian public health officials that recently arrived immigrants from Europe were the most likely to die from infection.[8] Because of its dynamic economy, European immigrants flocked to southern Brazil, avoiding the less-developed northern region.

The war not only slowed the rate of immigration, but it also served to undermine the already weakened interest in creating a comprehensive *mata-mosquito* program in Brazil's vulnerable coastal cities. However, as many foreign countries increasingly engaged in tropical commerce, their interest in yellow fever expanded. The United States displayed particular concern in this regard. In the wake of the Spanish-American War of 1898, expansion of U.S. political and economic hegemony into East Asian and Caribbean tropical regions was accompanied by a growing apprehension of diseases like yellow fever, malaria, and hookworm. Effective U.S. expansion into these areas required the adequate implementation of new biological techniques that minimized the costs of conquest, managerial settlement, and eventual incorporation of a given region into the U.S. economic and political sphere of influence. U.S. managers worried that chronic and epidemic disease added layers of unpredictability to the task of administering a multinational enterprise. To borrow from Robert Wiebe, the American search for order had gone global, and critical to the process was the development of cost-effective biological techniques to guard against the unfortunate economic side effects of specific endemic and epidemic diseases.

Drastic anti-yellow-fever and antimalaria programs requiring large-scale fumigation were an outgrowth of U.S. military operations in and around the Caribbean. The establishment of Cuba as a virtual U.S. protectorate and the spectacular success of building the Panama Canal[9] both utilized the talents of Army medical personnel like Generals Walter Reed[10] and William Gorgas,[11] among others. Though expanding, direct U.S. military presence in the world limited itself to regions vital to U.S. interest, like the Caribbean and East Asia. Global dominance of the world's oceans still eluded the ambitions of the U.S. Navy. Given the dimensions of the international yellow-fever problem, involving the three continents of Africa, North America, and South America, development of a comprehensive strategy against the disease necessitated the employment of an agency with much greater range and flexibility than the U.S. military could provide.

It is no accident that the breach was filled by the philanthropy of large, corporate, American capitalism. Philanthropic organizations like the Rockefeller Foundation performed a number of functions besides mollifying critics of the large trusts that men like John D. Rockefeller established in the second half of the nineteenth century. Superficially, these organizations soothed the bad conscience of men branded as "robber barons," by permitting fabulous sums of money to be spent in their name "toward the well-being of mankind."[12] But at the level of global management, U.S. big-business philanthropy, especially Rockefeller money, promoted improvements in the productivity of workers throughout the world as a means of enhancing the international division of labor between producers of raw materials and finished goods. Theoretically, productivity gains not only translated into increased raw-material output but also broadened the prospects of opening new consumer markets in regions of the globe that had been inaccessible to products of U.S. big business. Workers who produced more—even in isolated agricultural regions—were liable to buy more. To this extent, corporate philanthropy became a concern of U.S. foreign policy, facilitating U.S. political and economic hegemony over the underdeveloped world.

The Rockefeller Foundation, through its subsidiary International Health Board, had a particularly keen interest in Brazil. The largest of all Latin American countries, Brazil dominated the South American continent, meaning for all intents and purposes that any comprehensive strategy of disease management in the Americas that the Rockefeller Foundation planned required the incorporation of Brazil into the plans. By late 1916, General Gorgas, working on behalf of the foundation, was sent on a special mission to Brazil to investigate the possibilities of establishing a national anti-yellow-fever program.

Though the foundation suggested its participation in Brazil could effectively eliminate yellow fever from the country, Gorgas's arrival aroused the suspicions of certain sectors of Rio's intellectual elite. Placido Barbosa, who worked very closely with Oswaldo Cruz in the 1904 anti-yellow-fever campaign in Rio de Janeiro, argued the entrance of the Rockefeller Foundation into Brazil might be a kind of "advance guard" for other forms of imperialist domination. An editorial writer for the Rio newspaper, *Correio da Manha*, downplayed such fears, pointing out that a "new nationalistic orientation" was required that could affirm "the national personality" but at the same time initiate the "cooperation and solidarity of civilized peoples." The writer suggested a "new order of things" would follow the conclusion of the European war, an order in which a country trying to avoid internationalism "will perish like a member amputated from the body of collective civilization."[13] At least for the time being, debate subsided in Brazil over such matters, allowing the relatively smooth entry of the Rockefeller Foundation into the country.

Though yellow fever was an important concern, interest in other urban and rural diseases received as much (if not more) of the attention of the Rockefeller Foundation, at least in the foundation's early years in Brazil. In a report on public health conditions in Brazil written in 1920, the International Health Board underscored the prevalence of tuberculosis and syphilis in urban areas, and hookworm and malaria in the countryside.[14] Hookworm, in particular, received the attention and resources of the Rockefeller Foundation because of the facility of its treatment.[15] The foundation had already made a reputation for itself in its efforts to eliminate hookworm disease in the southern United States[16] and in other countries.

The growing international reputation of the foundation's hookworm program was not lost on Brazilian sanitary authorities, who recognized the potential benefits such a program might have on the productivity of rural workers, especially in the coffee industry.[17] Authorities from a number of Brazilian states and the federal government wasted little time in signing contracts with the foundation to establish hookworm-eradication programs. By the early 1920s, dozens of hookworm demonstration posts had been completed.[18] The paltry sums that Brazilian governmental agencies had set aside for rural sanitation—in 1917, about $12,000 was allocated for the entire country—had by 1922 increased to more than U.S. $2,000,000.[19]

The foundation, like its public counterparts in Brazil, focused priorities on diseases other than yellow fever. However, reports of minor outbreaks of yellow fever in a number of small towns in the northeast continued to surface in the early 1920s.[20] Brazilian public health officials took some solace in the fact that urban instances of the disease would almost certainly elicit a concerted political response, and thus give impetus to remobilizing the all-but-defunct *mata-mosquito* squads.

Cost-conscious officials of the Rockefeller Foundation, on the other hand, complained that dependence on outdated fumigation methods produced more dividends in the way of public relations than in public health. The foundation proposed that controlling mosquito breeding places was, "a simpler, less expensive, and more effective operation."[21] Less flamboyant and persistently more intrusive, antilarval work did not capture the imagination of local leadership or the general public in the same way that antimosquito work did. In his report on health conditions in Brazil, Wickliffe Rose of the Rockefeller Foundation contended that Brazilian public health officials used fumigation as a "show-piece" to hide management deficiencies which, he claimed, were endemic in all parts of government administration.[22]

In spite of such assumptions on the part of many Rockefeller Foundation personnel, Lewis W. Hackett, head of early operations in Brazil, found some comfort in a political situation he described as "kaleidoscopic." He noted that it "is consoling to consider that we can hardly be eclipsed by local factional disputes in all parts of such a large country as Brazil at the same moment."[23]

Under such circumstances, the degree of success in instituting certain changes in public health policy depended upon understanding the dynamics of local, regional, and national politics in Brazil, and, if need be, using such political knowledge as a way to gain a foothold in otherwise slippery terrain. Hackett, one of Harvard's first graduates of a special public health administration program, gained extensive experience in Guatemala, Honduras, and Panama during the war years.[24] In comparison to Brazil, however, these countries were virtual satellites of the United States, and imposition of new public health techniques there encountered, at best, token resistance. As Hackett counseled, Brazil would require a subtler approach.

Yet Rose's observations warrant further investigation. The Rockefeller Foundation, using techniques derived from experience in Central America and the Caribbean, disdained the Brazilian practice of killing adult *Aedes aegypti*. More cost-effective, from the Foundation's perspective, was to concentrate on eliminating *Aedes aegypti* in its larval form, thus terminating the life cycle before the insect had a chance to lay its eggs and infect human subjects. The near-deification of Oswaldo Cruz and his method of fumigation became an entrenched feature of anti-yellow-fever orthodoxy in most quarters of the Brazilian public health service. Rockefeller Foundation officials often lamented the adherence to such practices, deeming it impossible to organize a coordinated, national anti-yellow-fever program.

Nonetheless, the foundation quickly realized the important political function of fumigation: for the local public health officer, the demonstration effect of fumigation was an essential weapon in garnering the support of local political bosses and the cooperation of the public. Images of hostile public reactions to measures taken by a then-untested and undistinguished Oswaldo Cruz—including the calamitous Vaccine Riots of 1904—remained firmly entrenched in the minds of authorities. Though newer evidence suggests other sources of discontent contributed to the grave nature of the disturbance,[25] nevertheless, Brazilian health officials understood the danger of imposing ill-conceived measures on an urban population apprehensive of public power. The success of fumigation as a measure designed not only to eliminate yellow fever, but to soothe residents irritated by blood-hungry *Aedes aegypti* reinforced the links between medical faculties, public patronage, and local political bosses, ensuring the continued use of such measures, especially during election cycles.

For Brazilian health authorities, always under intense public scrutiny, deviating from orthodox practices meant putting their possible career advancement at risk. In cities where medical traditions were firmly entrenched, such as Rio and Salvador, and, to a lesser degree, Sao Paulo,[26] the implementation of newer methods was considerably more dangerous from a political standpoint than in the smaller cities of the northeast. In these smaller cities, like Recife, Fortaleza, Sao Luis, and Belém, medical culture did not penetrate as deeply and therefore did not maintain the same strong bonds with local polit-

ical machines. It was in these northeastern cities that the Rockefeller Foundation hoped to demonstrate the efficacy and economic efficiency of its methods,
prior to the eventual introduction of a unified program of yellow-fever control
that would encompass the entire country.

The foundation's biggest hurdle was convincing state and local public
health officials to dispense with methods considered politically correct, in favor
of more-efficient techniques that might have potentially damaging political
repercussions. Besides necessitating constant inspections, the antilarval approach practiced by the foundation tampered with, and in some cases destroyed, domestic water-storage systems. Standard practice required inspectors to drop small, larvae-eating fish or a layer of oil into water containers in
order to discourage mosquito breeding. Coastal urban centers in Brazil had
grown so fast during the previous thirty years that few major cities in the country had adequate underground water works available to serve residents. And
those that did, like Salvador, suffered from chronic disruptions in supply. This
compelled nearly every household to depend upon *jarras* (pitchers) or *caixas
d'agua* (water tanks) as primary or secondary sources of water storage. The fe-
cal matter of fish, dead fish, and oily residues added to the public's doubts
about the prudence of antilarval measures. The approach was intrusive, and
it had potentially damaging political repercussions for the foundation, especially when Brazilian physicians tied to the local political process remained un-
willing to cooperate.[27]

On a larger strategic level, there were problems as well. The Rockefeller
Foundation subscribed to a theoretical approach expounded by Henry Rose
Carter that drew a sharp distinction between large-city and small-town infections.[28] As Joseph H. White, director of the Rockefeller Foundation's yellow-
fever program in Brazil during the mid-1920s, wrote, "Our method of combatting yellow fever is very simple. We take for our field of operation only the
larger centers of population and pay no sort of attention to the small towns."[29]
Such an approach took for granted the urban nature of yellow fever. But subse-
quent findings in Brazil and elsewhere in Latin America would radically mod-
ify and expand the number of etiologic agents to include at least five other
species of mosquitoes, as well as certain monkeys and marmosets that har-
bored the disease in a sylvan or "jungle" form.[30] However, Rockefeller medical
personnel did not acknowledge the existence of other possible etiological cycles until the early 1930s.[31]

In the meantime, the efficacy of the foundation's narrow etiological focus,
concentrating on larger urban centers, did not take long to test. By the summer
of 1923, the yellow-fever situation in the northeast of Brazil had noticeably de-
teriorated. Informal negotiations begun by the Rockefeller Foundation in 1916
to take over large sectors of the Brazilian anti-yellow-fever service were con-

tractually formalized in the final months of 1923. Postwar reforms had already created a new federal Sanitary Code that had given officials in Rio de Janeiro the opportunity to intervene in public health affairs that were previously matters of state responsibility. This shift of public health authority away from the states in favor of the federal government underscored an ongoing trend of the 1920s; the decade witnessed a significant expansion of federal powers in a number of areas once deemed the prerogative of individual Brazilian states.

Accords were signed with eleven northern states, among them Bahia, Pernambuco, Alagoas, Paraíba, Rio Grande do Norte, Maranhao, and Pará.[32] Central authorities in Rio de Janeiro utilized newly created powers of federal intervention to give the Rockefeller Foundation permission to begin implementing its programs in a number of cities throughout the region. In November, A. W. Walcott, a Rockefeller representative, established an anti-yellow-fever post in Recife, where persistent cases of the disease had been reported. Similar posts were organized in all major ports from Belém to Salvador, including Manaus.

As the Rockefeller Foundation assumed a greater role in the campaign against yellow fever, conflict over methods and lines of authority surfaced in Salvador, a bastion of northeastern medical culture.[33] Sebastião Barroso, who since 1920 had been in charge of the federal government's anti-yellow-fever program in Bahia, resigned. His stated reasons were that the foundation had exceeded its limits of authority by assuming complete control in the anti-yellow-fever program. Barroso apparently assumed Rockefeller participation in the program would be more collaborative in nature. In fact, he charged, the takeover by the foundation had left him as a mere titular head without any real authority. Lacking control, Barroso was helpless to prevent the Rockefeller Foundation from implementing its antibreeding methods, which included the placement of fish in drinking water; Barroso claimed the fish used in this way were collected from streams heavily polluted by sewage. He opposed the practice of inspecting private residences for *Aedes* larvae and pupae, as well as the offering of prizes to district inspectors for the elimination of mosquitoes because, he claimed, incentives like these encouraged falsified reporting. The American consul who attended Barroso's public resignation, in front of the Bahian Medical Society, noted that "[q]uite a little nationalistic feeling was evident in the various speeches and no little resentment that foreigners were permitted to exercise authority in Brazil."[34] As if to reinforce the links between public health practice and nationalistic sentiment, the Medical Society adopted resolutions praising Barroso's work in Bahia and expressing regret at his resignation.

Of special concern to Barroso and other Brazilian public health officials was the Rockefeller Foundation's decision to focus efforts specifically in the larger coastal cities of Brazil, while virtually neglecting smaller hinterland towns. Barroso and his colleagues began to notice disturbing instances of rural yellow-fever infection. Drs. Waldemar Antunes and Clóvis Correia, representing the newly created federal Sanitary Commission in Pernambuco, discovered as early as 1920 what is believed to be the first diagnosed case of sylvan yellow fever in Brazil. Barroso noted similar cases coming from backland *fazendas* (large rural estates) in Bahia, and announced these findings at a conference in 1922.[35] What these cases suggested was that not only did yellow fever operate differently than Rockefeller representatives presumed, but the disease infected huge rural expanses of many of the states in which the foundation operated.

Resistant to the inferences that yellow fever might exist in a rural form, the Rockefeller Foundation successfully pushed forward its rigorous, urban anti-yellow-fever measures throughout its area of control. In Recife, for example, the city was subdivided into zones administered by vigilant teams of inspectors that visited anywhere from 200 to 700 structures per week, depending upon building density.[36] Such efforts produced impressive early results. Abel Tavares, the newly appointed head of the federal government's anti-yellow-fever service in Salvador, reported in 1924 that not one case of the disease had been observed in the city during the hot summer months. Like his predecessor, Sebastião Barroso, Tavares nevertheless cautioned that gains achieved in Salvador might be suddenly reversed by the "reimplantation" of the disease from neighboring states or infected parts of Bahia's interior.[37]

On the other hand, foundation administrator White, a veteran of a number of Central American campaigns, concurred with Tavares's conclusions on Salvador, but refused to acknowledge the possibilities of rural infection. He pointed out that 1924 was the first year in which yellow fever had not claimed a single victim in over half a century within the entire state of Bahia. From such evidence, he enthusiastically predicted the complete eradication of the disease from Brazil. White suggested that, with elimination imminent, the foundation would be able to concentrate personnel and resources in certain parts of West Africa where the disease reigned practically unimpeded.[38]

White's exuberance over his organization's achievements in the north and northeast of Brazil was shattered by events that unfolded during the early 1920s. A small-scale civil war, which had begun in the south, moved up through the *sertao* (backlands) of the northeast. Its causes were firmly rooted in the contradictions of a Brazilian political economy striving toward industrial development yet unable to extricate itself from the demands of the agricultural-export sector. The march turned into a wandering three-year expedition in the backlands of Brazil, which came to be known as the "Prestes Column," named for its leader Luís Carlos Prestes. Hoping to enlist the support of the common people residing in the countryside (where the majority of Brazilians still lived in the late 1920s), the column turned northward on a journey that would eventually carry it far into the state of Maranhao.[39]

Leaders of the insurrection clearly overestimated the support they could gather in the backlands. The rebel army, which in the latter stages of its existence fluctuated between 1,000 and 2,000 soldiers, was unable to convince rural workers to join in the uprising. In fact, many workers responded to the incantations of the threatened *fazendeiros* (rural estate owners) by taking up arms against the rebels. Facing repeated pitched battles against bands of armed workers known as *jagunços*, the rebels only on rare occasions received significant support in hinterland areas. Particularly nasty battles took place in the backlands of Bahia. In September of 1926, authorities in Rio de Janeiro, hoping to crush what remained of the tattered insurgent army, sent a contingent of troops from Sao Paulo numbering about 3,000 strong.[40] Though outright victory was never achieved by the government's *legalista* troops, the introduction of the regular army did initiate the full-scale retreat of the rebel force. In February of 1927, 620 would-be revolutionaries surrendered to Bolivian authorities, thus bringing to an end a convoluted twenty-five-month, 25,000-kilometer odyssey.

As for the anti-yellow-fever program initiated by the Rockefeller Foundation in the northeast, the backland civil war was the severest test of its urban strategy of coping with the disease. Almost immediately, the shortcomings of that approach became alarmingly apparent. The substantial number of nonimmune troops that had passed back and forth between focal points of infection in the hinterland *sertao* and the Bahian capital brought with it a series of officially diagnosed cases of yellow fever among the *legalista* troops.[41] As scores of soldiers, most of them from Sao Paulo state regiments, filled Salvador's *Hospital Militar*[42] physicians worried about the possibility of an outbreak occurring in the Bahian capital itself. Optimism, though, was high that the antilarval measures undertaken by the Rockefeller Foundation would limit the problem to the military's hospital facilities. Because the latest *Aedes aegypti* indices registered well below what foundation authorities believed to be optimum levels, containment of the outbreak was considered a likely possibility. "Honor and glory to the Rockefeller [Foundation]," exclaimed one local observer who publicly expressed his feelings for the work which had been conducted in Salvador and other port cities of the northeast.[43]

In spite of such public accolades, however, internal assessments of the situation were far more critical. A Brazilian physician, D. F. Freire, who worked for the federal government's yellow-fever commission, asserted in a letter to the Rockefeller Foundation's headquarters in New York that epidemic yellow fever had been in existence within the Bahian interior since the end of 1925— well before nonimmune government troops were sent to fight the insurgents.

Moreover, individual cases of the disease, according to Freire, had been inaccurately diagnosed as typhoid or malaria, with fatalities amounting to several hundred.[44]

How was it that U.S. experts in the field of yellow fever, with years of experience in the Caribbean and Central America misdiagnosed so many cases? Was there something different about the way yellow fever functioned in Brazil? M. E. Connor, who had replaced White as director of yellow-fever operations in Brazil, concluded in a letter to New York in 1928 that the Americans made errors "simply because the clinical entity did not conform to that which we had been accustomed to see in other parts of the Americas." He noted that previous yellow-fever experiences among Rockefeller Foundation personnel had been limited to "whites or near whites, but never *with blacks in large numbers and in whom the infection seems to assume, more frequently, the atypical form than in the whites and near whites*" (Connor's emphasis).[45] For over a year following the epidemic reintroduction of yellow fever into Salvador, the North Americans failed to recognize what many Brazilian physicians familiar with yellow fever already suspected: that many persons of African descent tended to display yellow-fever symptoms in much milder forms. F. F. Russell, director of operations in New York, acknowledged the insight to a U.S. colleague:

> Connor has now got to believe that this difference in the . . .
> composition of the population between Brazil and the rest of the
> American tropics is important. In Mexico, Central America, Colombia
> and Ecuador we had to deal mainly with the Indians and some
> Europeans and many mixed breeds, but with very few negroes and
> mulattos. In [the Northeast of Brazil] . . . are very few pure whites,
> many negroes and mulattos. . . . [M]ost cases are mild and difficult [to]
> diagnose.[46]

The shift in perceptions about the disease, however, did not occur until the second half of 1928. For nearly two years, Rockefeller Foundation personnel continued to misdiagnose cases coming from the interior of the northeast, and consequently underestimated the extent of infection in the region.

In the meantime, however, Rockefeller Foundation officials and federal authorities alike braced themselves for a possible outbreak. A major preoccupation was the Carnival of 1927. Rockefeller officials had suspected since 1925 that there was a link between Carnival and the disease.[47] What officials worried about most was the mass movements of population back and forth from the interior to the coast.[48] The situation in Salvador was particularly disturbing. By March, signs of trouble in the form of increased instances of the disease began to appear. Connor estimated that in addition to the city's population of 350,000, more than 50,000 persons moved through the port on an annual basis. Most disquieting was the existence of approximately 26,000 water tanks in addition to tens of thousands of other storage units which the population kept on hand to cope with the inadequacies of the city's waterworks.[49]

From a long-term standpoint, the availability of so many potential larval breeding places was thoroughly unacceptable. Yet it was neither politically nor logistically feasible to eliminate the massive number of water-storage units in Salvador, at least in the interim. For the time being, officials decided, the standard Rockefeller Foundation practice of dropping fish or oil in these containers would have to suffice. By late April, Connor proclaimed to Russell in New York that the antilarval campaign was proceeding with good results, "especially north of [Salvador,] Bahia, and not 'too bad' in that city." He estimated intensive measures would have to continue in Salvador, Recife, and Fortaleza until July 1929.[50]

In order to facilitate permanent measures that could ensure the eventual elimination of water-storage containers, officials in the Rockefeller Foundation attempted to persuade Brazilian authorities and the general public to invest in extensive urban waterworks projects. The U.S. firm of Ulen & Company, which had already installed systems in the Brazilian coastal cities of Porto Alegre and Sao Luis, utilized the Rockefeller Foundation in an effort to win contracts for future construction projects. Connor's concern for the water-delivery system in Salvador was so great that in early 1927 he mentioned to an Ulen representative "that the city might listen to a proposition for increasing [its] water system."[51] Similar meetings were held thereafter, with Connor acting as a go-between for Ulen and the state of Bahia. Connor even initiated a propaganda campaign "with the knowledge and consent of [Bahia's Minister of Public Health,] Barros Barreto, by pointing out to the householder that [he] would be relieved of weekly inspection[s] once . . . water is supplied in abundance."[52]

Negotiations bogged down, though, leading to continued disruptions in supply that forced householders to hoard and factories to close. Connor, somewhat surprisingly, reacted with delight: "[A]ll of this was in our favor for the louder the people howled about the scarcity of water, the sooner would the authorities be forced to take some definite action . . . to increase the quantity available to the public."[53] Yet none of this came about. Salvador's credit, which Connor described as "nil," required city authorities to ask for state and federal assistance in financing such a project.[54] Each of these levels of government placed enough restrictions upon the contracting agent that Ulen & Company decided the investment was not worth the risk. Eventually the city of Salvador contracted with a Rio-based engineering firm to renovate its waterworks.[55] However, construction was not scheduled to begin until the early 1930s. Toward the end of 1927, Connor lamented, "We must accept the idea that [yellow

Table 1: Yellow-Fever Mortality in Urban Rio de Janeiro, 1850–1909

YEAR	DEATHS	YEAR	DEATHS	YEAR	DEATHS
1850	4,160	1870	1,118	1890	719
1851	475	1871	8	1891	4,456
1852	1,943	1872	102	1892	4,312
1853	853	1873	3,659	1893	825
1854	22	1874	829	1894	4,852
1855	3	1875	1,292	1895	818
1856	101	1876	3,476	1896	2,929
1857	1,868	1877	282	1897	159
1858	1,545	1878	1,176	1898	1,078
1859	500	1879	974	1899	731
1860	1,249	1880	1,625	1900	344
1861	247	1881	257	1901	299
1862	12	1882	89	1902	984
1863	7	1883	1,608	1903	584
1864	5	1884	863	1904	48
1865	—	1885	445	1905	289
1866	—	1886	1,449	1906	42
1867	—	1887	137	1907	39
1868	3	1888	747	1908	4
1869	272	1889	2,156	1909	—
				60-Year Total:	59,049

Source: Graça Couto and Cassio de Rezende, *Control of Infectious Diseases in Brazil and Especially in Rio de Janeiro* (Rio de Janeiro: Typo. = Lithographia Pimenta de Mello, 1912), 41–42.

fever] is to be eradicated from Bahia without the benefit of a [modernized] water system."[56]

The distinct possibility that Salvador could quickly become an endemic center of yellow fever, from which other cities in Brazil would become infected, worried the Brazilians and the North Americans alike. Connor had already intimated to Russell that he believed the city of Salvador to be a "seed-bed" of infection.[57] The specter of a coastal eruption of the disease—not unlike the pattern of yellow-fever epidemics that savaged the country throughout the second half of the nineteenth century[58]—seemed hauntingly real once more (see Tables 1 and 2). Connor wondered why imported cases were not already apparent in Rio de Janeiro and Santos, and then speculated it "may be due to the anti-mosquito measures applied . . . by the National Health Department . . . which have reduced the degree of *aegypti* infestation." Falling short of endorsing fumigation, Connor considered both cities "infectible territory" because of the seasonal presence of *Aedes aegypti*.[59] Increasingly, Rockefeller Foundation officials feared a chance introduction of infection into the south from Bahia.

Table 2: Number of Cases of Yellow Fever in Urban Rio de Janeiro in 1928–29, Distributed by Month

MONTH	1928	1929	Total
January	—	29	29
February	—	54	54
March	—	241	241
April	—	190	190
May	4	87	91
June	52	9	61
July	40	1	41
August	9	—	9
September	10	2	12
October	2	—	2
November	2	—	2
December	6	—	6
Total	125	613	738

Note: Of the 738 officially reported cases of yellow fever in Rio, 438 resulted in death.
Source: Odair Franco, *História da Febre Amarella no Brasil* (Rio de Janeiro: Ministério da Sáude, Departamento Nacional de Endemias Rurais, 1969), 102.

Despite such concerns, federal authorities in Rio de Janeiro apparently were not fully apprised of the potential danger at hand. Antonio Luis de Barros Barreto, who simultaneously directed the Bahian state public health service and the federal government's yellow-fever program in Salvador, informed the head of the National Department of Public Health (DNSP) in Rio de Janeiro in late March 1927 that he was satisfied with the progress of the campaign. The federal government's top public health officer, Clementino Fraga, himself a former head of the Bahian public health service, had expressed concerns about possible infections in the backlands, where he believed yellow fever was endemic.[60] Nevertheless, both he and his predecessor, Carlos Chagas, severely reduced antimosquito measures in Rio beginning in 1926, under the assumption that the Rockefeller Foundation, in cooperation with state and federal authorities, had matters under complete control[61] and that the "fear of importation" had disappeared.[62] Connor later learned from unofficial sources that reductions in Rio's antimosquito work began taking place after July 1927, and that by January 1928 efforts of this kind had ceased altogether.[63]

In a report published in 1930, Fraga defended his actions by claiming that other diseases, such as leprosy and tuberculosis, demanded more of the attention and resources of the federal government's public powers. Furthermore, he asserted that domestic *Aedes aegypti* breeding areas were presumed to be relatively few in number, and the responsibility for those focal points of infestation was a problem which must be "transferred to householders" themselves. He complained that even if the political will had existed to reproduce the anti-mosquito infrastructure of Oswaldo Cruz's era, budget cuts had deprived the DNSP of the necessary means to carry through such a program. He argued that public perceptions of yellow fever in Rio de Janeiro, at the time prior to the outbreak, simply could not sustain an extremely costly antimosquito campaign against "a disease that did not exist."

Fraga concluded by stating that many in the capital believed the reimportation of the disease from the north was "quite improbable," especially considering the optimism that those charged with administering the northern campaign had earlier declared.[64] Fraga's inference was clear: because of White's publicly stated optimism regarding the program's successes in the northeast, the Rockefeller Foundation bore a large measure of responsibility for leaving Rio de Janeiro vulnerable to a yellow-fever outbreak. A more cautious approach on the part of foundation officials might have alerted Rio's public health authorities to the potential dangers at hand.

By May 1928, the rift between the Rockefeller Foundation and the DNSP over methods and control had clearly intensified. As rumors circulated in Rio about the presence of yellow fever in the capital, the need to coordinate strategies between the northern and southern sectors of the country assumed ever greater importance. Connor, writing to Russell from Bahia, expressed concern that the campaign stood at a crossroads. The foundation had to decide whether it was prepared to expand vastly its scale of operation, "to include all communities of 2,500 inhabitants and upwards in our inspection." Admitting the ambitiousness of such a proposal, Connor specified that an "enlarged program calls for a very heavy expenditure of funds and must embrace the areas from and including Rio to the Amazon valley and from the coast to well into the interior." He acknowledged the fact that the large-city strategy "has not been sufficient to effect the eradication of yellow fever from Brasil." For the time being, however, Connor recommended continuation of the same general strategies, but he added that the "takeover [of] Rio control measures" was a possibility.[65]

Connor's ambitions in Rio included the implementation of the same types of antilarval measures instituted in northern coastal regions of Brazil. Although the Rockefeller Foundation was able to overcome initial resistance in Salvador, Rio offered impediments of a more profound nature. The city was, after all, the nation's capital. Like Salvador, it enjoyed a long tradition of medical culture that extended back into the early nineteenth century. Local medical practitioners hired by the government had engaged foreign interests in conflict over methods of coping with yellow fever since the initial outbreak of the disease in 1849.[66] Despite repeatedly failing to come up with an effective strategy for blunting the damage caused by the disease, the Rio medical community, by practical default, claimed yellow fever as an object of investigation and experimentation.[67]

In the realm of yellow fever, Rio's medical community achieved its crowning glory during Oswaldo Cruz's tenure as head of public health services. Despite hostile resistance from old-guard medical practitioners,[68] Cruz and his medical-research laboratory, Manguinhos,[69] ably established legitimacy for Brazilian applied scientific endeavors during the early years of the twentieth century by successfully eliminating focal points of yellow fever infection using fumigation methods. Cruz's medical triumph coincided with the radical redevelopment plans of Rio's mayor, Francisco Pereira Passos.[70] Physicians had long argued for urban restructuring as a means toward mitigating the bad health effects for which Rio had become infamous.

The efforts of Oswaldo Cruz to eliminate yellow fever were an important contribution to what historians refer to as the *Belle Époche* of Rio.[71] The medical profession embraced his success as not only a breakthrough in the yellow-fever problem, but as demonstrative proof that Brazilian medical science possessed the capabilities to overcome tropical "defects" previously thought insurmountable.[72] Serving as a bulwark against the social Darwinistic tendencies of the era,[73] the application of Finlay's principals by Cruz and his colleagues challenged national and international assumptions that explained Brazil's backwardness in terms of inherently racial and climatic factors.[74]

Foreign incursions into the terrain of anti-yellow-fever technique in the nation's capital connoted breaches in not only the realm of tropical science, but the very sovereignty of the nation itself. As the congressional debates of 1928 demonstrate, the memory of Cruz's achievement in Rio de Janeiro remained firmly entrenched in the cultural consciousness of Brazil's national political elite. In striking contrast, Connor's cavalier attitude of "taking over" Rio's yellow-fever operation failed to account for the issues of national pride, which the reintroduction of the disease in 1927 raised.

Once it was evident that indeed yellow fever had returned to the capital—on May 23, 1928, a soldier who had recently arrived from the infected northeastern region was officially diagnosed with the disease[75]—the national debate began in earnest. Throughout the halls of the National Congress, deputies evoked the name and legacy of Oswaldo Cruz, suggesting that success or failure would be measured in the terms that Cruz had established in 1904. Amaury de Medeiros, a former state public health official and a deputy from Pernambuco, advised restraint in criticizing Fraga's management of the capital's public health situation. He cited the initial difficulties from which Cruz suffered, and suggested that indiscriminate criticism of Fraga would unnecessarily impair efforts to rectify the situation.[76]

The fragmented oppositional factions in the congress,[77] however, could not resist the temptation of attacking the national government's highest public health representative. Inverting de Madeiros's rhetorical tactics, the opposition repeatedly questioned Fraga's competence by comparing his efforts to Oswaldo Cruz's achievements.[78] The opposition intensified the debate by proposing emergency legislation to allocate additional funds to combat the disease. The set of proposals was overwhelmingly defeated because the governing majority that supported President Washington Luis and his "coffee mandate" did not wish to legitimize any claim that the national government had decreasing competence in the management of the capital, let alone the nation.

The legislative efforts of the opposition did, however, initiate a debate on matters of national public health policy. In spite of the government's first impulse to brand anyone who questioned Fraga's policies as "unpatriotic,"[79] by June 1928, congressmen as well as many people on the street wondered, if there were cases of yellow fever appearing in Rio during the middle of winter, what would things be like when the hot season began later in the year? No less concerned were officials of the Rockefeller Foundation who considered possible strategic alternatives. Connor's first impulse was to continue efforts toward an eventual Rio takeover. Though such a proposal was "discreetly" suggested to Fraga, Connor wrote to Russell that an invitation to begin antilarval measures "has not been forthcoming, and I doubt if it will in the very near future." Nevertheless, Connor predicted that Rockefeller assumption of control was only a matter of time, perhaps a few months. He was already developing plans for just such a contingency.[80]

However, the takeover of Rio's anti-yellow-fever operation had a major drawback that greatly concerned Rockefeller Foundation officials: caustic attacks began to surface in Rio's daily newspapers, drawing attention to the apparent mismanagement of the yellow-fever problem. Some of the attacks even criticized the work of the Rockefeller Foundation. Both Connor and his colleague, Fred Soper, agreed that the Rockefeller Foundation should maintain a low profile, at least for the time being. A Brazilian physician, Tomas Alves, who worked closely with the foundation, suggested calling on President Washington Luís as a way to increase Rockefeller influence in the matter. Connor resisted, stating that the foundation's work had no reason to defend itself.[81]

Connor, nevertheless, remained convinced that success in subduing yellow fever would come only with a unified program that maintained direct Rockefeller Foundation leadership in Rio de Janeiro. When told by Alves that there were men working against this objective inside the DNSP, Connor replied that "if Fraga did not desire our cooperation I was certain that our Division would try to force it upon the country."[82] Connor did not specify how that was to take place. Clearly, Connor underestimated the importance of yellow fever from a nationalistic point of view. A number of congressmen were already venting their ire at the steps initiated by foreign countries—especially neighboring Argentina—to restrict trade with Brazil until the yellow-fever problem had been sufficiently addressed.[83] With international attention increasingly focused on Rio's plight, pressure intensified to rectify the situation. Yet to seek direct Rockefeller assistance in the nation's capital—even under emergency circumstances—was to admit incompetence in a field where Brazilians once held some mastery. For Fraga, it certainly would have meant the end of his career as head of the National Department of Public Health.

On the other hand, there was also a crucial advantage for the Rockefeller Foundation in maintaining a low profile in Rio de Janeiro: as long as Fraga remained in charge, *he* became the issue of the campaign while the Foundation had the opportunity to work behind the scenes. Soper pointed out to Russell that, while the congressional debates subjected Fraga to intense criticism, the foundation "was left pretty much out of the discussion."[84] The inference was clear: a Rockefeller takeover would subject the foundation to all, if not more, of the pressure that Fraga absorbed as head of the nation's public health department. Considering there was no guarantee of success—Rio's population had more than doubled since the time of Cruz, leaving a considerable nonimmune population at risk—perhaps it would be more prudent to institute technical changes through Fraga's good offices.

Clementino Fraga did not passively operate as a marionette for the Rockefeller Foundation. Though faced with the severest crisis of his career as a public health officer, not only did he recognize immediately the opportunity to challenge a legion of hostile critics who surfaced at the time,[85] but, more important, he saw the chance to impress his medical-community peers in light of the achievements of his mentor, Oswaldo Cruz.[86] Success in such an effort also held the possibility of influencing national political leaders, who then might be inclined to extend Fraga's national authority. Already, the federal government was engaged in a trend to take over a number of health responsibilities previously delegated to the states, especially in the areas of rural health.

Even with desires on the part of the Rockefeller Foundation to dominate a Rio antilarval eradication program, Connor recognized Fraga's participation would be crucial, if only to ensure the cooperation of certain sectors of the general public. Pushing too hard would certainly alienate Fraga, thus turning one of the campaign's most important allies into a likely obstructionist. His contacts with Rio's medical community were particularly important since no plan could work without the cooperation of Rio's physicians. One of Connor's first recommendations to Fraga following the outbreak was to encourage a newspaper propaganda campaign that would use friendly journalists to persuade sectors of the general public of the efficacy of antilarval work.[87]

A combination of minimum antilarval measures, intensification of fumigation, and cooler weather produced reductions in the incidence of yellow fever in Rio de Janeiro. So successful had work been that the attacks against Fraga, which characterized the early outbreak period of May 1928, were significantly diminished by August (see Table 2). To Fraga's relief, he saw the successful control of yellow fever in Rio as proof that the methods of Oswaldo Cruz still produced impressive results. Extending such logic to the entire country merited use of these methods in the geographical domain of the Rockefeller Foundation, which included practically everything north of Rio de Janeiro. Soper attributed Fraga's fumigation ambitions to the flattery he received from Rio's medical community.[88]

Fraga did not stop there. In renegotiations concerning a new yellow-fever contract between the Rockefeller Foundation and the federal government, the head of the DNSP insisted on a reduction of foundation authority in those areas where the foundation remained, such as Salvador. Under the new agreement, the North Americans would come under Fraga's direct control. Noticeably irritated, Connor wrote, "Plan amounts to our supplying funds without control. My recommendation is to continue for present and consider withdrawal at end of year if situation remains the same."[89] Fraga's counterproposal, which Connor considered more moderate, still had clauses giving Brazilian officials authority equal to that of the foundation's representatives. Fearing misunderstandings would arise—which, in the final analysis, would be settled by Fraga—Connor expressed that he had no objections to this particular stipulation, "except that the psychology of the average Brazilian official in the [DNSP] is such that he feels superior to all others in yellow fever matters." Extended DNSP authority necessarily implied a national expansion of Rio's fumigation measures, which prompted Connor to comment that Fraga probably could be successful if he were "given unlimited funds and could pursue a consistent and persistent effort, but here is the 'rub'—the work would be carried on only while cases were occurring."[90]

As negotiations continued, the Rockefeller Foundation worried that the absence of yellow-fever cases would make its position more difficult in relation to the DNSP. Soper speculated that it "may be necessary to deal directly with higher authorities on this matter," as Alves had suggested previously.[91] Soper intimated to Connor that he believed several cases were occurring in Rio that the DNSP did not see fit to publicize.[92] By late September, cases that were once the object of speculation became official. Under such circumstances, the problem of reaching an agreement with the Rockefeller Foundation was overcome, leading Russell in New York to comment that the arrangements were not entirely satisfactory, but were "reasonably so."[93]

The agreement implied consent on the part of the Rockefeller Foundation for the fumigation methods of the DNSP. Fraga had indicated to Connor his decreasing reliance on such measures, preferring to use his personnel and resources for antilarval work. Nevertheless, Fraga, in Connor's view, was compelled to yield "to the clamor of worshippers of Oswaldo Cruz's methods."[94] Conceding the use of such measures, Connor began to study their application more closely. Fraga dispensed with Cruz's sulphur fumigant, using instead a mixture of kerosene, carbon tetrachloride, and pyrethrum, which was applied with a Du Pont spray pump. A major producer and importer of the types of petrochemicals used in this process was Standard Oil, the underwriter of the Rockefeller Foundation. In fact, nearly all the leading magazines and journals of the era published catchy advertisements demonstrating the effectiveness of "Flit," a commercially available antimosquito insecticide produced and distributed by Standard Oil of Brazil.[95] Yet there seems to be no evidence which suggests the Rockefeller Foundation's representatives in Brazil openly promoted the use of insecticides produced by Standard Oil, at least in yellow-fever work. To the contrary, the foundation immediately recognized the temporary nature of fumigation. That, coupled with its expense and inefficiency, caused Rockefeller administrators to agree to its use only reluctantly and after recognizing fumigation's important political function.

Nonetheless, the insistence of the DNSP on maintaining and improving fumigant antimosquito techniques[96] focused Rockefeller attention on the possibilities of future wide-scale insecticide applications in Brazil. Connor even admitted that, "theoretically, this insecticide may be developed as an auxiliary measure in reducing adult mosquitoes" in places where populations were otherwise out of control.[97] More than anything else, the change in policy on the part of the Rockefeller Foundation's director of yellow-fever operations in Brazil was recognition for the political importance of such measures. Urban politics in Brazil during the 1920s compelled public officials to take actions producing immediate observable results. Killing adult mosquitoes with elaborate fumigant devices had such an effect, whereas the antilarval method seemed more passive and long-term in its applications.

The reassessment of fumigation measures even compelled Bahia's minister of public health, Barros Barreto, to request the foundation to assume such measures in Salvador. Connor wrote,

> This situation will be tense for a time and we must go carefully. Soper was ready to agree to our fumigating. . . . I am not. If we start this sort of work, . . . our field of usefulness [will become] seriously hampered.[98]

Connor's apprehensions were so acute that he considered withdrawing the Rockefeller Foundation rather than using its resources to ends that he believed addressed political rather than epidemiological issues.

Fraga, meanwhile, privately admitted to Soper the long-term importance of antilarval work. The head of the DNSP recognized final credit for yellow-fever eradication in Brazil would be directly linked to the antilarval service.[99] Fraga's efforts in Rio de Janeiro demonstrated his confidence in such methods. Using his local Rio contacts, Fraga promoted the organization of the Cruzada de Cooperaçao na Extinçao da Febre Amarella (Crusade of Cooperation for the Eradication of Yellow Fever, or CCEFA). A number of important representa-tives of the press, industry, and high society filled its ranks.[100] Publications like the widely distributed *Revista da Semana* and the chic weekly *Fon-Fon* dedicated small, but conspicuously placed sections to "the combat of yellow fever."[101] The methods suggested by CCEFA were very much in concert with Rockefeller antilarval measures. "In a gutter, in a vase with flowers, in a can, in a broken bottle . . . the mosquito can leave its eggs," read an excerpt from one advertise-ment of the organization.[102] The appeal was not limited to the literate middle and upper classes. Fox Film and the power utility Light coproduced a propa-ganda film demonstrating domestic antilarval measures that residents could undertake. The film was exhibited in open-air venues throughout Rio and its suburbs, and CCEFA officials estimated that perhaps 50,000 viewers attended screenings.[103]

So effective were Fraga's efforts in Rio that Soper, more than a little dis-mayed, in July 1929 wrote to Russell in New York:

> I was surprised after a year of working for and planning on unified
> yellow fever service for all Brazil to learn that Dr. Connor, previously
> the strongest advocate of such [a] unified service, no longer believes
> such unification necessary.[104]

Connor had been impressed with Fraga's commitment to antilarval work, which placed special emphasis upon inspection and the adoption of fish larval control by the DNSP. Connor commented, "While in Rio I received information from sources regarded as reliable which encourages me . . . that . . . Fraga is now emphasizing our methods, thus effecting a unified cooperative cam-paign." A number of local physicians confided in Connor that fumigation was considered a very secondary measure, with its use continuing for "political rea-sons."[105] In Connor's judgment, methods used by the DNSP and the Rockefel-ler Foundation had achieved de facto unification.

The elimination of yellow fever from Rio de Janeiro and other important coastal cities in Brazil—a growing reality by the early 1930s—came down to a question of mosquitoes, namely *Aedes aegypti* mosquitoes. All concerned par-ties, both Brazilian and North American, had no doubts about such a premise. Where divergence did occur was in the realm of method, and it was here that the Rockefeller Foundation learned the political value of mosquitoes in Brazil. In undertaking a crucial role in the anti-yellow-fever campaign of the late 1920s, the foundation became increasingly aware that Brazilian public health administrators, usually representing state governments, used elaborate fumigation methods as a way of making a political impression upon local constituents.

To break the grip that state public health officers had in defining proper anti-yellow-fever strategies and procedures, the Rockefeller Foundation and the DNSP settled upon the antilarval approach as the method of choice in combatting the disease. In conflicts that developed between the DNSP and various state governments such as Rio de Janeiro (state) and Bahia, Fraga's organization was able to assert greater authority in the promotion of foundation methods.[106] The efficacy of such methods, clearly apparent by early 1929, deprived state public health authorities of an opportunity to demonstrate prowess in the field of yellow-fever work. In turn, the DNSP's takeover of such a project greatly depoliticized efforts to eradicate *Aedes aegypti*, thus allowing for the creation of Brazil's first nationally coordinated, bureaucratically conceived, anti-yellow-fever organization.

Fraga's decision to implement antilarval methods succeeded because the federal government did not feel the immediacy of political repercussions as acutely as state officials did. The insulation of the federal government, however, did not extend into the federal district of Rio de Janeiro itself, where an apprehensive constituency with a good deal of historical memory of Oswaldo Cruz's triumphs openly questioned the methods imposed by Fraga. As a token gesture, small squads of *mata-mosquitos* still operated the Du Pont pumps, but on a greatly reduced scale. Fraga's main concern was to apply systematic antilarval measures in as extensive an area in and around the capital as was possible. The large-scale geographic requirements of the project practically ensured that during Rio's hot summer of 1929 cases of yellow fever were bound to occur. And, indeed, they did (see Table 2). But the most striking difference between this particular outbreak and the one that had occurred six months previously was the support the Rockefeller Foundation bestowed upon Fraga. At Fraga's request, the foundation chose to suppress information about the outbreak, thus abetting Fraga's attempt to cover up the situation.[107] As the crisis passed and antilarval measures were instituted in virtually every corner of Rio de Janeiro, an alliance of methods was forged and bonded between the DNSP and the Rockefeller Foundation.

One recent writer on the subject of public health in Brazil during the First Republic, Luiz Antonio de Castro Santos, argues that "the contribution of the Rockefeller Foundation . . . [to the Brazilian public health movement] . . . should not be overestimated."[108] He cites a favorable political climate that mitigated the potential of a nationalist backlash against the organization. Yet it is clear that in the struggle between state and federal public health authorities—a supporting theme of de Castro's study—the Rockefeller Foundation helped to break the hold that state officials had on yellow-fever work. In this respect, the Rockefeller Foundation contributed to a process of legitimizing federal power in the realm of public health work. The foundation also created important anti-larval organizations, which the federal government was able to slip on like tai-lor-made gloves designed for urban yellow-fever work.

As important as Rockefeller Foundation inputs were in the campaign, it should not be overlooked that the U.S. organization was the recipient of a good many Brazilian contributions in the field of yellow-fever research. Despite their early resistance, foundation personnel finally began to accept the idea—first proposed by Brazilian physicians working for the federal government—that yellow fever existed in Brazil in a sylvan or rural form. A crucial shift, the ac-ceptance of this new paradigm forced the foundation to confront the yellow-fever problem on a massively enlarged scale that included small towns as a central part of antilarval treatment.[109] Such an acknowledgment served to re-inforce further the alliance with Brazilian national public health authorities, since there was no other entity in Brazil capable of mobilizing the proper de-gree of resources, personnel, and information necessary to carry through such an ambitious undertaking.

Yet the Rockefeller Foundation's experience in Brazil revealed a sobering side effect: though it was possible to *control* urban yellow fever through effec-tive means, its total *eradication* as a disease—a dream long held by the founda-tion—was simply out of the question. It was one thing to organize antilarval measures in urban space, but quite another to undertake such a task through-out the vast uninhabited areas of interior Brazil—much of which was infected with sylvan yellow fever. The successful control of urban yellow fever during the late 1920s, ironically, underscored for the Rockefeller Foundation the limi-tations of its public health work.

Notes

1. For a thought-provoking essay on the reconceptualization of yellow fever which challenges traditional interpretations, see François Delaporte, *The History of Yellow Fever: An Essay on the Birth of Tropical Medicine* (Cambridge, Mass.: MIT Press, 1991).

2. Odair Franco, *História da Febre-Amarela no Brasil* (Rio de Janeiro: Ministério da Sáude, Departamento Nacional de Endemias Rurais, 1969), 91–94; Nancy Stepan, *Beginnings of Bra-*

zilian Science: Oswaldo Cruz, Medical Research and Policy, 1890–1920 (New York: Science History Publications, 1976), 113–14.

3. M. E. Connor, Diary, March 26, 1928, 38, Rockefeller Foundation Archive (hereafter RFA), R.G. 1.1, Series 305, Box 46, Folder 232, Rockefeller Archive Center (hereafter RAC).

4. Among a multitude of diseases—both endemic and epidemic—that afflicted residents of Brazilian cities, the most destructive were tuberculosis, gastroenteritis, typhoid, diphtheria, smallpox, measles, and a host of venereal diseases. See Sam Adamo, "Order and Progress for Some—Death and Disease for Others: Living Conditions of Nonwhites in Rio de Janeiro, 1890–1940," *Studies in the Social Sciences* 25 (1986): 23–26.

5. Luiz Antonio de Castro Santos, "Power, Ideology and Public Health in Brazil, 1889–1930" (Ph.D. dissertation, Harvard University, 1987), 139–50.

6. See Placido Barbosa, "Pequena História da Febre Amarela no Brazil," *Archivos de Hygiene* 3 (May 1929): 1; and Franco, *História*, 95.

7. *Jornal do Comercio*, October 25, 1916. Found in "No 7204, Comment of the Press and of the Director-General of Public Health of Brazil Concerning the Arrival of the Yellow Fever Commission in Rio De Janeiro," RFA, R.G. 5, International Health Board/Division, Series 2, Box 24, Folder 142, RAC.

8. Doctors and other interested parties noticed European-immigrant vulnerability right from the outset. See Donald B. Cooper, "Brazil's Long Fight against Epidemic Disease, 1849–1917, with Special Emphasis on Yellow Fever," *Bulletin of the New York Academy of Medicine* 51 (1975): 678–79.

9. Joseph A. LePrince and A. J. Orenstein, with an introduction by L. O. Howard, *Mosquito Control in Panama; the Eradication of Malaria and Yellow Fever in Cuba and Panama* (New York and London: Putnam, 1916); W. P. Chamberlain, *Twenty-Five Years of American Medical Activity on the Isthmus of Panama: A Triumph of Preventive Medicine* (Mt. Hope, Canal Zone: Panama Canal Press, 1929).

10. Howard Kelly, *Walter Reed and Yellow Fever* (Baltimore: Medical Standard Book Co., 1906).

11. For the types of measures employed, see William Crawford Gorgas, *Sanitation in Panama* (New York: Appleton, 1918).

12. Robert Shaplen, *Toward the Well-Being of Mankind: Fifty Years of the Rockefeller Foundation* (New York: Doubleday & Company, 1964).

13. "Exaggerated Susceptibilities," *Correio da Manha*, October 24, 1916. Found in "No 7204, Comment of the Press and of the Director-General of Public Health of Brazil Concerning the Arrival of the Yellow Fever Commission in Rio de Janeiro," RFA, R.G. 5, International Health Board/Division, Series 2, Box 24, Folder 142, RAC.

14. Wickliffe Rose, "No 7502, Observations on Public Health Situation and Work of the International Health Board in Brazil," RFA, R.G. 5, International Health Board/Division, Series 2, Box 25, Folder 153, RAC.

15. In a letter to Wickliffe Rose, dated May 6, 1918, Lewis W. Hackett of the Rockefeller Foundation wrote, "The hookworm problem seems to appeal to [the governments of Brazil] as a field of action which gives a certain guarantee of success. Our work in Brazil and particularly in the Federal District has received considerable publicity of a favorable nature." RFA, R.G. 1.1, Series 305, Box 15, Folder 134, RAC.

16. John Ettling, *The Germ of Laziness: Rockefeller Philanthropy and Public Health in the New South* (Cambridge: Harvard University Press, 1981).

17. See William W. Cort, William A. Riley, and George C. Payne, "Investigations on the Control of Hookworm Disease. A Study of the Relation of Coffee Cultivation to the Spread of Hookworm Disease," *American Journal of Hygiene* 3 (July 1923): 111–27.

18. Roy F. Nash, "Conquering the Hookworm in Brazil," *Current History* (March 1923): 1021.

19. The Rockefeller Foundation, *Annual Report, 1921* (New York: Rockefeller Foundation, 1921), 123–24. For more on the Rockefeller Foundation's hookworm program in Brazil, see

Steven C. Williams, "The International Health Board and Changing Urban/Rural Relations in Brazil," *Research Reports from the Rockefeller Archive Center* (Spring 1990): 13–15.

20. Franco, *História*, 96; "Dr. Borges to Dr. Noguchi," RFA, R.G. 5, International Health Board/Division, Series 2, Box 24, Folder 142, RAC.

21. Rose, "No 7502," 26.

22. Ibid.

23. Hackett to Rose, February 22, 1918, RFA, R.G. 5, International Health Board/Division, Series 305, Box 15, Folder 134, RAC.

24. Roy F. Nash, "Selling Public Health in Brazil: Five Years' Work of the International Health Board," *Brazilian-American* 5 (March 1922): 44.

25. See Teresa Meade, " 'Civilizing Rio de Janeiro': The Public Health Campaign and the Riot of 1904," *Journal of Social History* 20 (1986): 301–22; Jeffrey Needell, "The *Revolta Contra Vacina* of 1904: The Revolt against 'Modernization' in *Belle-Epoque* Rio de Janeiro," *Hispanic-American Historical Review* 67 (1987): 234–69.

26. For a comparison of the medical traditions of Sao Paulo and Salvador, see de Castro Santos, "Power," chapters 4 and 5.

27. M. E. Connor, Diary, February 25, 1929, 5.

28. Henry Carter later clarified the point in the classic work on the subject, his posthumously published, *Yellow Fever: An Epidemiological and Historical Study of Its Place of Origin* (Baltimore: Williams and Wilkins, 1931).

29. Joseph H. White, "Memorandum Descriptive of Method of Work against Yellow Fever, 1925," RFA, R.G. 5, International Health Board/Division, Series 305, Box 23, Folder 138, RAC.

30. Carrol L. Birch, "Jungle Yellow Fever," in Thomas G. Hull, comp., ed., *Diseases Transmitted from Animals to Man* (Springfield, Ill.: Charles C. Thomas, 1963), 893–94; Loring Whitman, "The Arthropod Vectors of Yellow Fever," in George K. Strode, ed., *Yellow Fever* (New York: McGraw-Hill, 1951), 231–98.

31. Fred L. Soper et al., "Yellow Fever without *Aedes aegypti*. A Study of a Rural Epidemic in the Valle do Channaan, Espirito Santo, Brazil, 1932," *American Journal of Hygiene* 18 (1933): 555–87.

32. De Castro Santos, "Power," 145–46.

33. The depth of resistance to new medical methods in Bahia can be seen by its reluctance to institute Oswaldo Cruz's measures. Not until the 1910s did Bahian sanitary policy abandon its pre-vector-theory approach to yellow fever, which included disinfection and total isolation, in favor of Cruz's antimosquito fumigation methods. See de Castro Santos, "Power," 281.

34. Homer Brett, "Anti-Yellow Fever Work in Bahia by the Rockefeller Foundation," December 11, 1923, RFA, R.G. 5, International Health Board/Division, Series 2, Box 24, Folder 142, RAC.

35. Franco, *História*, 96.

36. Ibid., 105.

37. *Gazeta Médica da Bahia* (April 1924). Cited in Otto Schmidt, *A Febre Amarella na Bahia em 1926* (Salvador, Bahia: Livraria e Typ. do Commercio, 1926), 33–34.

38. See "Febre Amarela no Norte do Paiz," *Revista de Sáude Pública* (April 1925): 66–68.

39. Nelson Werneck Sodré, *A Coluna Prestes: Análise e Depoimentos* (Rio de Janeiro: Civilização Brasileira, 1978), 36–45; Neill Macaulay, *A Coluna Prestes* (Rio de Janeiro: Difel, 1977).

40. Macaulay, *A Coluna*, 222.

41. Joseph H. White, "Yellow Fever—Brazil: A Survey, 1926," RFA, R.G. 5, International Health Board/Division, Series 2, Box 23, File 138, 2, RAC; Putnam Peris, "Yellow Fever in Brazil, 1926," RFA, R.G. 2, International Health Board, Series 305, Box 550, File 3673, RAC.

42. White, "Yellow Fever—Brazil," 2.

43. Professor Garcez Fróes, "Discurso de Saudação a Noguchi na Faculdade de Medicina da Bahia," cited in Schmidt, "A Febre Amarella," 193.

44. D. F. Freire to G. Jameson Carr, October 4, 1926, RFA, R.G. 5, International Health Board/Division, Series 2, Box 23, Folder 158, RAC.

45. Connor to Russell, July 20, 1928, RFA, R.G. 5, International Health Board/ Division, Series 305, Box 46, File 232, RAC.

46. Russell to Simon Flexner, August 10, 1928, RFA, R.G. 1.1, Series 305, Box 20, Folder 159, RAC.

47. In a letter to White, G. Jameson Carr compared the damaging effects of Brazil's Carnival to Mexico's Revolution of 1910. March 23, 1925, RFA, R.G. 5, International Health Board/ Division, Series 2, Box 26, Folder 156, RAC.

48. Connor, Diary, February 9, 1927, 48.

49. Ibid., March 12, 1927, 72–75.

50. Connor to Russell, April 20, 1927, RFA, R.G. 1.1, Series 305, Box 19, Folder 155, RAC.

51. Connor, Diary, February 18, 1927, 54.

52. Ibid., May 17, 1929, 65.

53. Ibid., October 9, 1927, 183.

54. Ibid., May 15, 1927, 109. F. F. Russell commented to M. E. Connor, in a letter dated June 22, 1928, that the state of Bahia had authorized U.S. $5 million for Salvador's future waterworks. RFA, R.G. 1.1, Series 305, Box 20, Folder 158, RAC.

55. Connor, Diary, July 26, 1929, 113.

56. Ibid., September 28, 1927, 180.

57. Ibid., March 12, 1927, 72–75.

58. Franco, *História*, 24–44.

59. Connor, Diary, May 12, 1927, 72–75.

60. Ibid., March 30, 1927, 85.

61. Ibid., September 25, 1929, 138.

62. Clementino Fraga et al., *A Febre Amarella no Brasil: Notas e Documentos de uma Grande Campanha Sanitária* (Rio de Janeiro: Off. Graph. da Insp. de Demographia Sanitária, 1930), 8.

63. Connor, Diary, September 9, 1928, 164.

64. Fraga et al., *A Febre Amarella*, 8.

65. Connor to Russell, May 21, 1928, RFA, R.G. 1.1, Series 305, Box 20, Folder 158, RAC.

66. Especially active were the British, who feared that Brazil might impose a rigid *cordon sanitaire* that would impede Britain's lucrative commercial interest in the country. At the onset of the first epidemic, the British delegation went so far as to issue a series of consulate dispatches that tried to convince Corte officials in Rio de Janeiro that the origin of the disease was not its importation on commercial ships—as many concerned Brazilian physicians believed—but rather local focal points of insalubrity. The British were quite adamant in pointing out the futility of standard quarantine practices, suggesting instead that yellow fever could be more effectively managed through sanitary policies that placed special emphasis on cleaning certain dockside flophouses that the British considered areas of infection. See Johns Westwood's letter to Joaquim Maria Nascentes d'Azambuja, August 23, 1853, call number 8–3–7, Arquivo Geral da Cidade do Rio de Janeiro.

67. The author is currently completing a doctoral dissertation which explores the galvanizing effect—both from a medical and a social perspective—which yellow fever had on Rio's medical community during the nineteenth century.

68. Stepan, *Beginnings*, 89–91.

69. Jaime L. Benchimol, coor., *Manguinhos: Do Sonho à Vida* (Rio de Janeiro: Casa de Oswaldo Cruz, 1990), 5–88; Stepan, *Beginnings*, 97–100.

70. Jaime L. Benchimol, *Pereira Passos: Um Haussmann Tropical: A Renovação Urbana da Cidade do Rio de Janeiro no Início do Século XX* (Rio de Janeiro: Biblioteca Carioca, 1990), 204–75.

71. Jeffrey D. Needell, *A Tropical Belle Epoque: Elite Culture and Society in Turn-of-the-Century Rio de Janeiro* (New York: Cambridge University Press, 1987).

72. Stepan demonstrates the crucial link between science and medicine in her book. See *Beginnings*, 56–83.

73. See Thomas Skidmore, *Black into White: Race and Nationality in Brazilian Thought* (New York: Oxford University Press, 1974), 51–53.

74. Afránio Peixoto complained bitterly in an article widely disseminated outside of Brazil, "The Climate of Brazil," in *The Brazilian Year Book—1908* (New York: G. R. Fairbanks, 1908), 15–21, that Europeans placed too much emphasis on climate as a factor in "man's development." For a discussion of these ideas, see Donald B. Cooper, "Oswaldo Cruz and the Impact of Yellow Fever on Brazilian History," *Bulletin of the Tulane University Medical Faculty* 26 (1967): 52.

75. Fraga et al., *A Febre Amarella*, 5.

76. *Díario do Congreso Nacional 28a Sessao, em 14 de Junho de 1928* (Rio de Janeiro: Imprensa Nacional, 1928), 704–705.

77. One of the most outspoken leaders of the opposition was Joao Batista Luzardo, a medical doctor from Rio de Grande do Sul who had organized anti-bubonic-plague measures in the city of Uruguaiana in the late 1910s. His participation in the 1923 revolution and later contacts with Luís Carlos Prestes gained him a reputation as a rabble-rouser within the halls of the Congress. See *Dicionário Histórico-Biográfico Brasileiro: 1930–1983*, vol. 4 (Rio de Janeiro: Editora Forense-Universitária, 1984), 1965–1966. See also Macaulay, *A Coluna*, 156.

78. *Díario do Congresso Nacional 53a Sessão, em 20 de Julho de 1928* (Rio de Janeiro: Imprensa Nacional, 1928), 1577.

79. The issue of patriotism was raised at the outset of the 1928 congressional debates on the subject of yellow fever. *Diario do Congresso Nacional 28a Sessão*, 712.

80. Connor to Russell, June 6, 1928, RFA, R.G. 1.1, Series 305, Box 20, Folder 158, RAC.

81. Connor, Diary, June 7, 1928, 104.

82. Ibid.

83. *Díario do Congresso Nacional 53a Sessão* (Rio de Janeiro: Imprensa Nacional, 1928), 1545, 1575–76, 1578, 1642–43. Brazil threatened to retaliate, claiming Argentine ports were infected with plague; see Fred L. Soper to F. F. Russell, July 22, 1928, RFA, R.G. 1.1, Series 305, Box 20, Folder 159, RAC.

84. Soper to Russell, July 22, 1928.

85. The most vicious attacks came from the sensationalist press. Newspaper mogul Geraldo Rocha owned three dailies that each took extremely aggressive positions against Fraga. Even the more middle-class journals criticized Fraga, although he was not without his supporters. See "A Febre Amarella e a Imprensa de Geraldo Rocha," *O Jornal*, March 7, 1929, 6.

86. Later in his career, Fraga attempted to insure his ties with the Cruz legacy by publishing a laudatory biography, *Vida e Obra de Osvaldo Cruz* (Rio de Janeiro: Livraria José Olympio Editora, 1972).

87. Connor to Fraga, June 7, 1928, RFA, R.G. 1.1, Series 395, Box 20, Folder 158, RAC.

88. Connor, Diary, September 16, 1928, 158.

89. Ibid., September 19, 1928, 161.

90. Ibid., September 21, 1928, 164.

91. Soper to Connor, August 28, 1928, RFA, R.G. 1.1, Series 305, Box 20, Folder 159, RAC.

92. Connor, Diary, September 16, 1928, 158.

93. Russell to Flexner, September 24, 1928, RFA, R.G. 1.1, Series 305, Box 20, Folder 159, RAC.

94. Connor, Diary, September 26, 1928, 169. Connor to Russell, July 18, 1929, RFA, R.G. 1.1, Series 305, Box 20, Folder 161, RAC.

95. The *Revista da Semana*, for example, regularly published advertisements for Flit. One, entitled "Um Inimigo Implacavel—O Mosquito," pointed out the health risks of mosquito

bites. The use of the product could protect the home from "this enemy which attacks at night." The copy claimed Flit to be a product perfected by chemists of "world renown." See *Revista da Semana* 29 (October 13, 1928): 4.

96. Joao de Barros Barreto and Antonio Gonçalvez Peryassú, "Da Aspersao de Insectisides na Prophylaxia da Febre Amarella," *Archivos de Hygiene* 3 (1929): 405–25.

97. Connor, Diary, February 23, 1929, 3.

98. Ibid., March 4, 1929, 7.

99. Soper to Connor, June 5, 1929, RFA, R.G. 1.1, Series 305, Box 20, Folder 161, RAC.

100. The organization leaders included Oscar Weinshenk, Jeronyma Mesquita, and Roberto Shalders.

101. *Revista da Semana* 30 (April 20, 1929): 31.

102. "A Febre Amarella: Suggestoes da CCEFA," *Revista da Semana*, 30 (May 4, 1929): 23.

103. "Guerra ao Mosquito!" *Revista da Semana* 30 (July 20, 1929): 30.

104. Soper to Russell, July 6, 1929, RFA, R.G. 1.1, Series 305, Box 20, Folder 161, RAC.

105. Connor to Russell, July 18, 1929, RFA, R.G. 1.1, Series 305, Box 20, Folder 161, RAC.

106. In one particularly acrimonious exchange between the DNSP's Fraga and Bahia's Barro Barreto, Connor wrote that Barros Barreto was ready to "Go to [the] floor," with Fraga over yellow-fever policy. See Connor, Diary, March 4, 1929, 7; Fraga's interference in the yellow-fever policies of the state of Rio de Janeiro prompted its Minister of Public Health, Alcides Lintz, to protest by temporarily leaving his post. See Soper to Russell, June 5, 1929, RFA, R.G. 1.1, Series 305, Box 20, Folder 161, RAC.

107. Soper reported that "many cases going to the isolation hospital have two diagnoses, one that is put on the hospital books for public and newspaper consumption, and one which is sent directly to the DNSP. . . . From the above items you can see how unwelcome would be an attempt on our part at this time to gather any real information on the local outbreak of yellow fever." Soper to Russell, March 11, 1929, RFA, R.G. 1.1, Series 305, Box 20, Folder 160, RAC.

108. De Castro Santos, "Power," 136.

109. Soper, "Yellow Fever without *Aedes aegypti*."

3

The Rockefeller Foundation in Revolutionary Mexico

Yellow Fever in Yucatan and Veracruz

Armando Solorzano

Present methodological concerns rightly stress the need to scrutinize the Rockefeller Foundation's programs according to a country's economic, political, and social conditions.[1] No attention has been paid, however, to analyzing such conditions on a scale smaller than that of a nation. The underlying assumption seems to be that the development of a nation is even and uniform; internal variation is rarely, if ever, considered. This chapter addresses this gap in the literature by studying the role the Rockefeller Foundation played in the campaign against yellow fever in two very different Mexican provinces, Veracruz and Yucatan. Veracruz is located on the Gulf of Mexico, about 270 miles east of Mexico City; Yucatan, about 940 miles from Mexico City, is one of the states that separates the Gulf of Mexico from the Caribbean Sea.

Consideration of these provinces is particularly interesting because the foundation's interventions there occurred during a political upheaval, the Mexican Revolution, and because the areas reflected an uneven spread of U.S. capitalism, different revolutionary experiences, and different attitudes toward the RF's campaigns on the part of the Mexican medical elites. In addition, this study considers Mexico's early-twentieth-century efforts to improve the health of the population, and the emergence of a national public health apparatus.

When the Rockefeller Foundation attempted to begin operations in Mexico in 1911, the nation was experiencing its most agitated and violent period. Mexico was in the early process of forming itself as a state, and lacked the resources to intervene effectively in sanitation. The peculiar conditions under which the foundation had to work informed the interactions between U.S. philanthropy and U.S. capitalism in Mexico during the early 1920s.

In fact, my work supports the hypothesis that the objectives and techniques of the RF's yellow-fever campaign were determined by the level of U.S. economic investment in the area and by the political conditions of the Mexican Revolution. Further, the campaign modified the revolution in significant ways: by transforming the anti-U.S. sentiments of the people of Veracruz, by helping stabilize and legitimate Mexico as a state, and by creating the basis for influencing future institutional developments in medicine and public health in postrevolutionary Mexico.

Veracruz, the United States, and the Rockefeller Foundation

By 1910 Veracruz was one of the most important U.S. economic enclaves in Latin America.[2] U.S. capital investments in the area were significant, and agricultural and industrial production was geared toward satisfying the demands of North American markets.[3] The monopoly of the oil industry, and the continuous U.S. acquisition of land through concessions, transformed the local peasantry from landowners to wage earners. U.S. investors in Veracruz saw the Mexican Revolution as a serious threat to their interests, a perception that prompted a U.S. invasion of Veracruz of 1914. In that year, under the pretext of "protecting" the life and property of U.S. citizens, the U.S. Marines temporarily occupied Veracruz, following an incident involving the arrest of U.S. sailors.

The incident intensified the anti-U.S. sentiment of a populace already chafing against U.S. political and economic control. The revolutionary aim of "Mexico for the Mexicans" found its strongest support in the nationalism that developed in Veracruz. When the U.S. military withdrew from the city, President Venustiano Carranza occupied it and made it his revolutionary headquarters. Local residents and politicians backed Carranza and his nationalist agenda of recovering Veracruz for the Mexicans. He made special efforts to reduce nationalism to a position of anti-Americanism, however, since U.S. investors controlled the mines, oil wells, and agricultural production of the region he was trying to control, and he was concerned that regionalism could become an obstacle to the efforts of the national government.[4]

The incident in Veracruz was one of the main reasons for the RF's delay in beginning operations in Mexico. The foundation had been interested in working in Mexico since the outset of the Mexican Revolution, and had attempted, without success, to start public health campaigns there from 1911 until 1920. Three factors impeded RF efforts: President Carranza's strong opposition, grounded in nationalism; the anti-American sentiments generated by the U.S. invasion of Mexican territory in 1914; and the Mexican government's general distrust of U.S. organizations. Political conditions began to change in the late 1910s and early 1920s when, after years of civil war, revolutionary leaders began to reconstruct the country, seeking stability, improved relations with the United States, and economic growth. A new president, Alvaro Obregón, came into power late in 1920, and in that same year signed an agreement with the RF's trustees for a campaign against yellow fever in the Mexican territory.[5] The agreement established that the program would concentrate in Veracruz.

The pervasive anti-U.S. attitude that prevailed in Veracruz, however, interfered with the Rockefeller Foundation's intention to inaugurate a campaign in the province. Wickliffe Rose, director of the International Health Board, was cognizant of the attitudes of the people at Veracruz. Moreover, he acknowledged that the Rockefeller name had been linked to one of the groups dominating the economy of the area:

> It is believed by many people in Mexico that the present political troubles are the result of outside economic interests; that prominent among these interests are the oil interests; and that Standard Oil has played an important part in it. Any organization therefore bearing the Rockefeller name would not be kindly received by the present government authorities.[6]

Yellow fever had existed in Veracruz since the nineteenth century, and the area was considered one of the world's major endemic centers of yellow fever. In 1903 the infection reached epidemic proportion, and the health authorities of Veracruz, directed by the distinguished Mexican physician Eduardo Liceaga, mounted an anti-yellow-fever campaign organized around the guidelines of modern sanitation and strict control of the vectors transmitting the disease.[7] The outcomes were outstanding; by 1910 the fever was controlled. Yet, as Table 1 demonstrates, the yellow fever returned forcefully in 1920.

A major difference between the RF campaign and prior Mexican efforts against yellow fever was that the foundation set a new goal: the eradication, rather than control, of the disease.[8] The task of eradicating the fever from Veracruz turned out to be one of gigantic proportions, especially because of the initial opposition of the people to any U.S.-directed sanitation measures.

Nourishing the revolutionary turmoil in Veracruz were, on the one hand, opposition toward the United States, and on the other, rejection of Alvaro Obregón as president. In 1920, anti-U.S. sentiment led the authorities of Veracruz to reject U.S. President Wilson's offer of $50,000 to improve the city's sanitary conditions and combat bubonic plague.[9] Opposition to Obregón and his pro-U.S. attitudes was overwhelming in Veracruz, as was support for Carranza, the deposed president. The peasantry of Veracruz saw Obregón as a serious step backward from the Revolution.[10] The foundation was not only

Table 1. Yellow Fever in Veracruz, 1900–1924

Year	Cases Reported	Deaths
1900	543	261
1901	269	103
1902	678	285
1903	1075	228
1904	73	12
1905	54	23
1906	17	12
1907	2	1
1908	43	20
1909	3	3
1910	0	0
1911	0	0
1912	0	0
1913	0	0
1914	0	0
1915	0	0
1916	0	0
1917	0	0
1918	0	0
1919	0	0
1920	485	235
1921	75	31
1922	20	11
1923	0	0
1924	0	0

Sources: "Yellow Fever Reports, 1922," RFA, R.G. 5, Series 3, Box 147, RAC; and Archivo Histórico de Salubridad, Folder 27. Data from 1900 to 1912 were taken from Francisco Castillo Najera, "Campaña Contra la Fiebre Amarilla," *Revista Médica Veracruzana* 1 (1921): 188–205.

subject to anti-U.S. sentiments, it was perceived as a collaborator of President Obregón's.

Moreover, the work of the foundation was perceived as preparation for a future invasion of the U.S. Army.[11] For example, at Papantla, movies on the Armenian massacres by the Turks were used to "show" what would happen in Veracruz if U.S. intervention would take place.[12] Thus, the Veracruz authorities and people "never cooperated" with the foundation,[13] and when the foundation's doctors arrived in some locations of Veracruz, the people persecuted them.

Although official statements suggest that the RF misinterpreted, or put aside the people's rejection, internal documents show the foundation's awareness of and deep concern with the attitudes adopted by the people of Veracruz. The following is just an example of the difficult conditions that the RF had to face there. In the city of Tuxpan, the RF personnel faced life-threatening conditions:

> In Tuxpan we were merely tolerated. . . . [O]ur Delegado Sanitario received absolutely no assistance from either officials or populace. Some of the medical men of the town refused to allow the inspectors to enter their homes, and even the President Municipal (a position corresponding to that of Mayor), refused to allow his water containers to be inspected. Some of the inspectors were driven from houses by machetes and pistols, while others were allowed to make inspections, but were subjected to more or less insulting remarks by residents.[14]

The merchants of Tuxpan showed similar attitudes, putting forward their own ideas concerning yellow fever. In a metaphor with political overtones, the merchants declared:

> The fever that devastates us is not just yellow but golden. It is not the stegomyia vector that produces the virus, but the Oil Companies and the institutions working with them: not in the form of mosquitoes but in the form of dollars. . . . It is a strange fever that can be immunized with GOLD, because there is no doubt that our COUSINS (North Americans), the ones who live beyond the [Rio] Bravo in The [Oil Station] Barra, the ones who travel by boat and car without any restrictions, those ones are immunized.[15]

Even the physicians of Veracruz rebuffed the foundation. Dr. Casasus and Dr. Loyo, Mexican doctors in charge of the anti-yellow-fever campaign in Veracruz prior to the foundation's presence there, did not approve of the foundation's plans. They claimed that the RF doctors knew no more about yellow fever than the Mexicans did. The differences were perceived to be the RF doctors' superior economic capacity and the way they treated local physicians:

> They do not waste time either making budgets or waiting for authorizations, they dispose of immediate funds and consequently they put to work all the people they need. . . . They look at us with a little disdain, they do not take our opinions into consideration, we are inferior to them; and although they treat us right most of the time, deep down you can see their desire to dominate, to impose on us always.[16]

Further complicating the environment in which the foundation had work was the conflict between the people of Veracruz and President Obregón. The peasantry in particular, under the leadership of Adalberto Tejeda, was demanding from Obregón the fulfillment of the promises of the Revolution of 1910.[17] Tejeda believed that President Obregón was not faithful to the spirit of the revolution. Although the revolution had brought some gains to the peasantry, the conditions of the peasants in Veracruz remained the same as, if not worse than, before the revolution. Thus, the people were disillusioned not only with the United States but also with Obregón's revolutionary government.

Yet for the RF representatives, the revolt in Veracruz was a "new type of revolution," substantially different from the revolution of 1910. It was not a revolt in which the peasantry represented the driving force; rather it was a conflict of power among different regional elites. The RF saw it as a "local revolution," lacking national implications.

By the end of 1922, however, the RF had come to fear civil war, and that fear became a reality. The state of Veracruz openly declared its opposition to Obregón. To the foundation's surprise, some of its own personnel took an active role in organizing the opposition to the national government.[18] Having the largest concentration of health workers in Veracruz, and having built a huge organization of public health in the state, the foundation believed it natural "for the Mexican Government to become suspicious of our most innocent activities."[19] Wishing to avoid endangering its relationship with the government, when the RF's doctors detected workers engaged in the revolt, the foundation dismissed these workers.[20] In addition Emmett J. Vaughn, director of the RF's anti-yellow-fever campaign, advised to Dr. F. F. Russell in New York to move part of the foundation's offices and laboratories from Veracruz to Mexico City.[21] The relocation had a twofold intention. First, the foundation wished to show its loyalty to President Obregón, and second, the foundation was trying to avoid the linking of its activities with the revolutionary movement in Veracruz.

As these events suggest, the RF's goal was not merely to gain the favor of the people for its campaign but also to support President Obregón, to whom the foundation had promised its solidarity and support. To overcome the opposition of the people to the U.S. and to Obregón, the foundation's doctors would have to convince the local government of their altruistic intentions.

Without doubt, Obregón needed to gain political control of Veracruz. Its resources were rich, and at the international level, it projected a negative image of Mexico as a whole. The reemergence of yellow fever in Veracruz was blurring the image of "modern Mexico" as a "civilized country," an image Obregón was trying hard to foster in the international arena. Promoting this image was especially important at a time when the government faced international distrust. Lacking sufficient financial resources and expertise, Obregón asked for the Rockefeller Foundation's assistance in the control of yellow fever in Veracruz, and the RF started its campaign in April of 1921.

Table 2. Yellow Fever Activities, 1921–1923

	1921	1922	1923
Houses inspected	883,031	2,186,026	2,575,869
Breeding places destroyed	94,210	151,903	52,249
Receptacles inspected	2,135,680	3,774,302	8,204,138
Liters of oil used	22,090	208,731	299,195
Fish deposited	160,054	1,013,787	3,368,111

Note: Data only include Tamaulipos and Veracruz.
Source: "Yellow Fever Reports, 1920–1921, 1922, 1923," RFA, Series 3, Box 147, 148, RAC.

The Campaign against Yellow Fever in Veracruz

The consensus among RF and many other medical experts was that the yellow fever could be eradicated by controlling the mosquito *Aedes aegypti*, which transmitted the fever and bred mainly in domestic water containers.[22] During the antilarval campaign in Veracruz, RF personnel used "simple but efficient" methods: house inspections, deposit of small fishes into water receptacles, and the use of petroleum in ponds and marshes. Yellow-fever brigades, each composed of one sanitary inspector, one supervisor, two assistants, and seven laborers, carried out the campaign, destroying the mosquitoes and larvae wherever they were found.

The most important part of the campaign was the screening of people's houses: "The instructions given each Inspector have been to enter and inspect each home for water containers. . . . "[23] Each house inspected was carefully reexamined every six days. To insure that all houses were visited and all the territory of the area was covered, the brigades held conferences every week.

By December of 1921, the number of yellow-fever cases had been reduced by 85 percent, and by 1923, when the fever was "eradicated" from Mexico, the RF brigades had carried out a monumental and exhaustive task. Table 2 summarizes the activities of the RF in Veracruz. In comparison with 1921, the number of houses inspected in 1922 increased 248 percent, and 291 percent in 1923. The reason for these increments was the foundation's policy of combatting yellow fever inside people's houses. Yet, visiting people's houses, with the intention of destroying the *Stegomyia*, was a method that played a fundamental role in reversing the anti-U.S. sentiments of the people.

By the end of 1923, the foundation had visited at least 41 percent of the rural population of Veracruz. Visiting "all houses without exception" was a measure strongly emphasized by the foundation, and the delicate issue of how to enter people's houses was broadly discussed by the foundation directors and their personnel. According to specific instructions, it was "highly important that sanitary officers find a way to enter houses and premises without irritating the people, as their cooperation is essential and must be obtained even at the cost of some months delay to an effective beginning of the work."[24] Reaching the population was an important component of the campaign. One officer was convinced that "our public health work is going to become really effective only insofar as we can enlist the active and sympathetic cooperation of the masses of the people."[25]

The rationale for making personal contact with "the common people of the region" was the need to gain their cooperation, but this effort was also instrumental in transforming the opposition found in Veracruz. The values embedded in the RF's campaign, the changes it required in community organizations, the health improvement, and the dedication of the Rockefeller personnel were expected to create new attitudes toward the United States and the Mexican government within the population. By the end of 1922, the work of the foundation had, in fact, changed people's attitudes. The spirit of dedication on the part of RF workers and the efficiency of some of the methods they employed brought the population to cooperate with the foundation. The foundation's representative in Mexico wrote to the director of the International Health Board, "About one half of the Indians in this area (San Andres Tuxtlan, Veracruz) do not understand Spanish but they can see the improvement in an individual after treatment and that is most excellent propaganda."[26] In addition, the commercial benefits of the campaign were so obvious that the merchants of Veracruz had to acknowledge them:

> [The RF campaign] had saved several lives that without the campaign would have been lost, the campaign made possible a surplus population which constituted the most important part of the businesses; it guaranteed the commercial imports which was reflected in the state's income, it liberated the port of Veracruz of the harmful quarantines, and it liberated also the country from the inherent burdens implied in the rising costs of the goods.[27]

However, the change in the attitude of the population was not only a result of the techniques applied against yellow fever. Besides eliminating mosquitoes and larvae, RF personnel opened new sewers in the city and carried out a street-cleaning campaign in Veracruz. The work of the RF became visible to the people and acquired greater legitimation. It was this popular, visible approach that opened doors to the RF. The work of the foundation implied immediate, personal contact with the entire population. Thus, a field officer reported that,

> No matter how uneducated, the people speak of *La Fundacion Rockefeller* with a real affection, for they have been individually

protected from one of the most dreaded diseases by staff members who have been willing to undergo considerable personal hardship to bring yellow fever immunization to them.[28]

Once the epidemic of yellow fever was under control, the RF announced the withdrawal of its presence in Veracruz. The people of the province reacted by demanding that the Mexican government take over the campaign.[29] Pressure on the Mexican state for maintaining La Comision Especial Contra la Fiebre Amarilla organized by the RF came from the working class of Veracruz. Sugar-mill workers, peasants, and railroad workers sent letters to the Mexican Department of Public Health. Workers appealed to the state to fulfill the revolutionaries' promises of improving the welfare of the working class.[30] Workers demanded that the benefits of the revolution be shared with everybody, especially with the workers and peasants who had fueled the revolutionary process. It was because of the yellow-fever campaign that the people started demanding public health programs.

The Mexican state had reached a point where it had to take responsibility for the welfare and public health programs. The foundation let the Mexican president know it was aware of the pressure to which the Mexican authorities were subjected. To alleviate the pressure, and following its rule of keeping a low profile in the administration of programs, the foundation decided to do whatever was necessary to present the success of the anti-yellow-fever campaign as a triumph of the Mexican administration. The foundation decided to remove itself from the front line and appear as a servant of the Mexican administration.

As a result, the Mexican state gained legitimation in the eyes of the Veracruz population. In addition, the RF benefited substantially, in that future work in public health was made easier. According to an RF representative, "the confidence of the people is such that we can kill a member of the family with chenopodium, and the other members will demand that they continue to receive their treatments."[31]

In part, the goals of the foundation for Mexico had been achieved. When the anti-yellow-fever campaign was over, F. F. Russell, director of the International Health Board, wrote to the president of the Rockefeller Foundation, "I have heard that there has been a real change of feelings in this country . . . that before the popular feeling was pro-German and pro-English but that now while it cannot be called pro-U.S. it is inclined that way."[32]

In order to appreciate the magnitude of the transformation brought by the RF in Veracruz, it is necessary to compare it with the campaign against yellow fever that the RF implemented in Yucatan.

Yucatan, the Mexican Revolution, and the Rockefeller Foundation

While the Mexican Revolution showed its violent nature in Veracruz, the people in Yucatan remained aloof to the national revolt. According to Joseph M. Gilbert, the Yucatecan people did not engage in revolutionary violence since Yucatan, and particularly its capital city Merida, were enjoying a "golden age" in the production and sales of henequen.[33] Henequen, a yellowish, hard fiber obtained from agave leaves and used for binder twine, among other purposes, was in great demand in the international market during the early twentieth century. Further, notions of regionalism and separatism from central Mexico had been characteristic of Yucatecan people since the prerevolutionary period.[34] Yucatan's location, remote from central México, and the reluctance of its people to join the revolution influenced the work of the RF in that area.[35]

It took five years for the revolution of 1910 to arrive in Yucatan. It was introduced there in 1915 by General Salvador Alvarado, who landed in the peninsula with 6,000 troops.[36] Surprisingly enough, the Mexican Revolution experienced a fascinating twist in Yucatan. It was not a cry for land redistribution there, as in many regions of the country,[37] or a struggle among regional elites, as in Veracruz.[38] After establishing control of the peninsula of Yucatan, Alvarado turned the Mexican Revolution into a local version of a socialist movement. The immediate goals were the consolidation of a Socialist party, the protection of the workers' rights, popular education, health benefits, and women's rights. Alvarado and his associate Felipe Carrillo Puerto transformed the revolution into the most radical movement of twentieth-century Mexico.[39]

Official U.S. reactions to these developments came through the U.S. consul in Merida. The diplomat was torn by the idea that Yucatan would adopt the Bolshevik doctrine defended by Marxism. Alvarado and Carrillo Puerto's support of socialism was interpreted as anticapitalist and subsequently as anti-U.S. But the consular reaction was off target. In fact, in no other region of the Mexican republic were the sentiments of the people so favorable to the United States as in Yucatan. Loyalty to Mexico was more likely to be contested. The willingness of Yucatan to be separated from Mexico was expressed in 1914, when the Yucatecans petitioned to become part of the United States.[40] The wide-open enthusiasm of Yucatan for the United States brought President Carranza to consider Yucatecans as "anti-patriotic" and "pro-American."

Once Carrillo Puerto became a governor, socialist reforms were implemented. An important goal was to recover the identity of the Mayan culture, that of the main indigenous group of the region, which could be traced to pre-Columbian times. His desire to reach the population and its segregated com-

munities brought him to construct roads and make more accessible the cere-
monial ruins of Chichen Itza and Uxmal, a work accomplished with the collaboration of the Carnegie Foundation.

In spite of the wide sympathy toward the United States, the American-
owned businesses, International Harvester and the Peabody Co., which controlled 100 percent of the Yucatan henequen industry, felt little confidence in
Alvarado or in Governor Carrillo Puerto's administration. To guarantee its
properties, International Harvester demanded that the U.S. diplomats petition
for an intervention of the U.S. government in Yucatan.

The U.S. government was willing to protect the henequen producers, but
the protection offered was against the infection of yellow fever. Through the
Department of Public Health, the U.S. government was aware of the possibilities of spreading the fever to different ports of the United States through the
exports of henequen from Yucatan. Overall, Mexico's and particularly Yucatan's possible exportation of yellow fever to the United States was a continuous concern to the U.S. government. The Rockefeller Foundation was also very
much interested not in controlling but in eradicating the disease from Yucatan,
which was considered one of the oldest endemic centers of yellow fever on the
continent.[41]

A complete depiction of the endemicity of yellow fever in Yucatan is provided in Table 3. Prior to 1906, the mortality and morbidity produced by the
fever were attributed to the lack of an appropriate campaign. However, in 1907,
Dr. Eduardo Liceaga, who was then director of the Mexican health department,
carried out an impressive campaign against the fever. Between 1907 and 1918,
following the cyclical behavior of the disease, there was a decline in the number of yellow-fever cases. In addition, the Mexican Department of Public Health
showed a great deal of interest in controlling the fever through local campaigns.

In 1918 Wickliffe Rose met in Washington, D.C., with General Gorgas, the
Surgeon General of the United States, and other U.S. functionaries to discuss
the yellow-fever situation in Yucatan. The consensus of the forum was that
Merida was an endemic center of yellow-fever infection and a constant menace
to Cuba and other regions of North America.[42] No Rockefeller Foundation
campaign commenced, however, until 1920, when the RF reached an agreement with the Mexican government. Due to the fact that no cases of yellow
fever had been reported in the region since late 1919, the foundation decided
to concentrate on preventing the reintroduction of the disease, by "cleaning
up" the province.

For M. E. Connor, director of the foundation's anti-yellow-fever campaign
in Yucatan, one of the main goals of the public health campaign was to promote
a healthy labor force.[43] Among the RF campaign directors there was no doubt

Table 3. Yellow Fever in Yucatan, 1899–1922

Year	Cases Reported	Deaths
1899	10	3
1900	39	12
1901	113	47
1902	112	35
1903	310	105
1904	114	39
1905	8	4
1906	119	73
1907	13	2
1908	67	35
1909	39	22
1910	0	0
1911	50	26
1912	27	11
1913	4	4
1914	27	8
1915	1	1
1916	30	9
1917	6	2
1918	0	0
1919	48	21
1920	0	0
1921	0	0
1922	0	0

Source: T. C. Lyster, "Report and Covering Cases of Yellow Fever, 1900–1920," RFA, R.G. 5, Series 2, Folder 198, RAC.

of the direct relationship between the harvest of henequen in Yucatan by American-owned companies and the spread of the fever. When crops were abundant, there was a need to import workers from other states to harvest the henequen. And while the locals had developed some immunity to yellow fever through exposure to it, the newcomers had not, and suffered the attack of the fever.[44] To maintain uninterrupted production of henequen, it was necessary to eliminate yellow fever. When henequen production was lean and demanded a small labor force, this too had implications for yellow fever: labor was not imported, and yellow fever almost disappeared.

T. C. Lyster, an RF officer, former colonel in the U.S. Army Medical Corps, and former director of the 1919 RF anti-yellow-fever campaign in Central America, called a meeting with the U.S. consul at Merida. At the conference, the diplomat and the doctor concluded that a campaign against yellow fever in Yucatan was mandatory because the disease, first, kept out foreign capital, and, second, prevented the Mexican government from sending in federal troops in large numbers, as they would be quickly and easily overcome by yellow fever.[45]

The presence of the Mexican Army was desired by the U.S. consul and by RF doctors as a means to check the strong influence of "bolshevism."[46] In 1920, when Alvaro Obregón was elected President of Mexico, he decided to extend the authority of the national government and fight against the socialist revolution in Yucatan. To guarantee the task of the Mexican Army, the RF considered it urgent to protect the soldiers against the fever. The Yucatecans were considered a low priority since a great number of them were already immune because of prior attacks or exposure to yellow fever.[47]

The RF's campaign was instrumental in furthering Obregón's goals. The economic and political conditions of Yucatan were very favorable to the RF's carrying out a successful campaign there. For the directors of the foundation, the optimal conditions for starting a public health campaign in a foreign country were a positive attitude on the part of the government and the probability of cooperation by the people.[48] If Veracruz had been a nightmare for the RF, Yucatan would turn out to be a paradise.

The Campaign against Yellow Fever in Yucatan

No cases of yellow fever had been reported in Yucatan since December 1919, so the campaign the foundation initiated in February 1921 centered on prevention and the protection of the nonimmune.[49] The two main techniques implemented were antilarval measures and the use of Hideyo Noguchi's vaccine and other protective measures for the Mexican soldiers sent by the national government. The antilarval methods attempted to eliminate or greatly reduce the population of *Aedes* mosquitoes. The use of the Noguchi vaccine was recommended and urged for troops that went into areas where yellow fever had been experienced in the past.[50] However, the Noguchi vaccine was based on a scientific mistake and was discovered in the late 1920s to be completely useless; moreover, a new form of yellow fever, jungle yellow fever, was discovered.[51]

Based on the apparent success of the vaccine, the federal government and Dr. Connor agreed that, before soldiers were sent into the interior, they should be protected with Noguchi's inoculation. The same soldiers were requested to come back to their camps for further auscultation.[52] After they were vaccinated with Noguchi's serum, the troops were sent to Colima, another state that was opposing the revolution and, in addition, rejecting the help of the RF yellow-fever personnel.

Although credit for preventing the fever among the Mexican soldiers was mistakenly given to Noguchi's vaccine, other factors produced the outcome.[53] First, prior campaigns had eradicated the conditions that made yellow fever possible. Furthermore, the cleaning of domestic water tanks and especially the confinement of soldiers in so-called "concentration camps" protected the soldiers. The camps were specially designed to keep the mosquitoes and the illness away from the troops. In the camps, water receptacles—common breeding places for the *Aedes aegypti*—were strictly controlled, and all soldiers suspected of contracting the fever were immediately placed in isolation. In addition to these preventive measures, soldiers were subjected to medical examinations on a daily basis.

Another of the factors preventing the introduction of the fever among the soldiers was the RF's decision that the fever could not be fully eradicated without attention to the conditions of yellow fever affecting the general population. Drawing on experience gained in prior anti-yellow-fever campaigns in South America, RF doctors agreed that the use of small larva-eating fish was the best measure to apply.[54] Using the fish was cheaper than and as efficient as petrolization, and it did not disturb the population as much. According to an RF officer,

> Citizens especially those of the poorer classes, have been glad to help us in this work [the use of fish]. The majority of persons of the higher classes are in complete harmony with this campaign. . . . The results secured in keeping mosquitoes away from their homes have convinced them.[55]

In spite of the overwhelming support for the foundation, however, some opposition was reported, apparently among the upper classes.[56] To guarantee full implementation of the program, the health department of Yucatan circulated an open letter appealing to the moral duty of the people to cooperate with the RF campaign and announcing penalties for those who did not.[57]

The use of fish in Yucatan diminished the need for sanitary brigades. It is noticeable that, while the RF needed 203 inspectors in Veracruz, there were only 37 people involved in the campaign in Yucatan. With fewer personnel and at very low cost, the RF was achieving extraordinary success.[58]

One of the main features of the Yucatan campaign was the support and cooperation of the people. Given their pro-U.S. attitudes, the Yucatecans had no qualms about accepting the advice of the foundation's doctors. People opened their houses to the brigades and were willing to cooperate. Dr. Connor considered this cooperation as the most "consistent factor" in the success of the campaign.[59]

The cooperation extended to the active participation of schools, which were encouraged to take an active part in the campaign. Children also played an important role. While the girls "could have fun" searching for and killing mosquitoes, boys could help deposit the larva-eating fish in the containers at home. With this "powerful contingent" of teachers and children, Dr. Connor saw the feasibility of the "final day" when yellow fever would be "eradicated" from Mexico.[60]

If the voluntary cooperation of the people was overwhelming, the support of the medical profession at Yucatan was truly astonishing. Unlike their counterparts in Veracruz, the physicians in Yucatan were from the very beginning friendly, cooperative, and to certain extent admiring of the actions of the RF. For example, Pedro F. Rivas, a Yucatecan doctor who specialized in yellow fever, wrote the foundation that its work in Yucatan "increased my admiration for the Foundation and for the intelligent and unselfish men who belong to it."[61] The openness of the medical profession was attributed to the fact that a number of the leading doctors of Merida were trained abroad and quite knowledgeable about U.S. and European medicine.

The decisive collaboration of the Yucatan doctors appears especially clear in the case of Diego Hernandez. The concentration of power by this physician can be revealed by the fact that Dr. Hernandez was a prominent political figure, the director of the health department of Yucatan, a Mexican congressman, and the mayor of the city of Merida. Any contact Dr. Connor established with Yucatecan politicians or physicians was mediated by the presence of Dr. Hernandez, whom the Americans considered to be "all in one"; that is, he carried the power of the sanitary office and Merida politics and represented Yucatan at the federal level. Not surprisingly, the presence of the RF was hardly disputed in these three political arenas of Yucatan.[62] The RF trustees fully appreciated this close collaboration with the doctors of Yucatan and offered them fellowships to study at U.S. universities and at the Rockefeller Institute.[63]

The cooperation of the Yucatecan people, the collaboration of their medical professionals, and the support of the national government that characterized the RF's work against the reintroduction of yellow fever in Yucatan was presented as an example for the Mexican people to follow:

> In Mérida, the frequent conflicts in other parts of the country, coming from the indifference and apathy of the people toward complying with their hygienic regulations, are absent. In Mérida the people have reached the highest level of education, as is proven by the fact that the same inhabitants demand inspection of their water containers and complain to the local authorities when the inspectors do not fulfill some of their obligations.[64]

In 1924 the RF announced that yellow fever was completely eradicated in Mexico. The statement was not completely accurate, however, since yellow fever had been absent in Yucatan since before the initiation of RF work. The early control of the fever in Yucatan contrasted with the situation in Veracruz, where the fever was not eradicated until December 1922, and invites comparison of the two campaigns and their implications.

Conclusions

By analyzing the RF's experience in Veracruz and Yucatan, we can draw some connections between the anti-yellow-fever campaigns and the revolutionary situation in Mexico. Certainly, the revolution brought political and economic instability, threatening U.S. economic and political interests in Mexico. These issues—social unrest and the stability of the state—were the doors that allowed the Rockefeller Foundation to enter Mexico and the Mexican Revolution. The RF provided stability in Veracruz by reversing the anti-Obregón sentiments of the people and by making the state of Veracruz more tolerant of the presence of U.S. interests. In the case of Yucatan, the RF indirectly contributed to the pacification of the region by protecting soldiers against yellow fever. Once protected, they were able to advance into Yucatan and consolidate President Obregón's position. The presence of the army in Yucatan was necessary to control the labor organizations promoted by Carrillo Puerto's socialism. With troops in place, Obregón broke the barriers of Yucatan's traditional insularism, and his army was ready to eliminate the socialist revolutionary uprising in Yucatan.[65]

The RF acted as an important component in expanding the legitimation of the Mexican state. The Mexican state emerging from the 1910 revolution was not recognized by all the people as a legitimate institution. For example, newspapers in Veracruz were continuously criticizing the federal government for spending state resources to strengthen its military position or to organize its own political campaigns, and for lacking the commitment to improve the health conditions of the nation.[66]

The RF was vital in legitimizing the position of the new revolutionary Mexican state since the foundation's campaigns were instrumental in presenting the Mexican government as a motive force in the improvement of the health conditions of the population.[67] The control of yellow fever was presented and perceived as a major manifestation of the Mexican state's being congruent with the principles of the Mexican Revolution, and with the constitutional commitments it made in 1917, when it promised universal health care for Mexicans.

But the role of the RF in the Mexican Revolution did not end with its help in pacifying and stabilizing conditions in Mexico. Well before the Mexican state could achieve economic solvency, the foundation took on the role of financing public health programs. Consequently, the foundation was willing to supply the Mexican government with the necessary funds to carry out the anti-yellow-fever campaign as well as basic sanitation for some areas of the country. During the campaigns, the foundation avoided bringing the issue of financing to the Mexican state. The RF decided to provide "all the funds needed" and not to require matching sums.[68]

The RF campaigns also had implications for ongoing association between the Mexican state and U.S. medicine. By financing the yellow-fever campaign, the foundation was creating the basis for future U.S. influence on the development of the Mexican health care system. After the 1920s, the RF became the most important source of financing for and consultation about health care services in Mexico. In other words, the evolution of social institutions in postrevolutionary Mexico was going to be deeply affected by the power and resources of the Rockefeller Foundation.[69]

The public health campaigns of the RF not only altered the course of the Mexican Revolution, they shaped local attitudes toward the United States and the development of Mexican medicine and public health.

Notes

1. Edward H. Berman, *The Influence of the Carnegie, Ford, and Rockefeller Foundations on American Foreign Policy* (Albany: State University of New York Press, 1983).

2. See John M. Hart, *Revolutionary Mexico: The Coming and Process of the Mexican Revolution* (Berkeley: University of California Press, 1987).

3. Alberto J. Olvera et al., "La Estructura Económica y Social de Veracruz hacia 1930: Un Análisis Inicial," monograph, Centro de Investigaciones Históricas, Universidad Veracruzana, 1981.

4. Friedrich Katz, *The Secret War in Mexico: Europe, the United States, and the Mexican Revolution* (Chicago: University of Chicago Press, 1981), 255–56. Besides its U.S. connection, the city of Veracruz became the headquarters of various revolutionary leaders. Its strategic location on the Gulf of Mexico allowed revolutionaries to flee the country in case of emergency. More important, Veracruz offered revolutionary factions extraordinary opportunities to get into central Mexico. In short, during the decade of the 1910s, Veracruz was an important focus of insurrection, saturated with violence, international disputes, and local politics. See Leonardo Pasquel, *Carranza en Veracruz en 1915* (Mexico City: Editorial Citlatepetl, 1976).

5. The details of the campaign, its eradication of the fever, and its consequences are fully analyzed in my doctoral dissertation. Armando Solorzano, "The Rockefeller Foundation in Mexico: Nationalism, Public Health and Yellow Fever, 1911–1924" (Ph.D. dissertation, University of Wisconsin–Madison, 1990).

6. Memorandum of Dr. Rose's interview in Washington, D.C., September 22–23, 1919, Wickliffe Rose, Diary, Rockefeller Foundation Archives (hereafter RFA), R.G. 12, Rockefeller Archive Center (hereafter RAC).

7. Francisco Castillo Najera, "Campaña Contra la Fiebre Amarilla," *Revista Médica Veracruzana* 1 (December 1921): 188–205.

8. The RF considered eradication a priority since Veracruz was one of the last endemic centers of yellow fever remaining in the Americas. M. E. Connor, "Notes on Yellow Fever in Mexico," *American Journal of Tropical Medicine* 3 (March 1923).

9. Adrian Carranza, "Fiebre Amarilla. Junta Vecinal de Salubridad y Mejoras de Veracruz," October 12, 1920, Folder 12. See also, Octaviano Gonzalez, "La Peste Bubónica En Veracruz," September 26, 1920, Folder 20, Archivo Histórico de Salubridad (hereafter AHS: the Archivo Histórico de Salubridad is located in Mexico City).

10. Heather F. Salamini, *Agrarian Radicalism in Veracruz, 1920–1938* (Lincoln: University of Nebraska Press, 1978).

11. An American doctor reported "the rumor—that seemed to be generally believed—that we were doing anti-yellow fever work pending the arrival of American troops." T. C. Lyster to Dr. Wickliffe Rose, September 24, 1921, RFA, R.G. 5, Series 2, Box 33, RAC.

12. Ibid.

13. Letter from Gabriel M. Malda to the Governor of Veracruz, December 29, 1920, Folder 13, AHS.

14. Letter from Emmett J. Vaughn to Dr. Rose, September 24, 1921, RFA, R.G. 5, Series 2, Box 33, RAC.

15. Enrique Hoyos Ruiz, "La Fiebre Dorada, Veracruz, 13 Noviembre 1920," AHS.

16. Letter from Cuaron to Vasconcelos, October 8, 1922, Fiebre Amarilla, Folder 17, AHS.

17. The organization of peasant leagues and the opposition of Tejeda to Obregón are analyzed in Salamini, *Agrarian Radicalism.*

18. Emmett J. Vaughn to Dr. F. F. Russell, December 25, 1923, RFA, R.G. 5, Series 1.1, Box 74, Folder 1057, RAC.

19. Emmett J. Vaughn to Dr. Warren, January 7, 1923, RFA, R.G. 5, Series 1.2, Box 193, Folder 2475, RAC.

20. Emmett J. Vaughn to Dr. F. F. Russell, December 25, 1923, RFA, R.G. 5, Series 1.1, Box 74, Folder 1057, RAC.

21. Letter from Emmett J. Vaughn to Dr. Warren, January 7, 1923, RFA, R.G. 5, Series 1.2, Box 193, Folder 2475, RAC.

22. The first cases of fever in Veracruz in 1921 could be traced to people immigrating from other states and other countries. T. C. Lyster, "Report of Yellow Fever for the Year 1921," October 14, 1921, Folder 31, AHS. See also, Najera, "Campaña," 183–205.

23. H. B. Richardson, "Yellow Fever Campaign in the Canton of Papantla," February 15, 1922, RFA, R.G. 5, Series 3, Box 147, RAC.

24. J. H. White, "Memorandum on Yellow Fever as an Eradicable Disease, July 30, 1914," RFA, R.G. 5, Series 2, Box 41, Folder 52, RAC.

25. Wickliffe Rose, memorandum, March 15, 1916, RFA, R.G. 3, Series 908, Box 15, Folder 151, RAC.

26. Warren to F. F. Russell, May 17, 1926, RFA, R.G. 5, Series 1.2, Box 258, Folder 3281, RAC.

27. Letters from merchants and owners of hotels to the president of the health department, October 5, 1921, Folder 14, AHS.

28. John C. Burgher, interoffice correspondence, October 25, 1950, RFA, R.G. 3, Series 908, Box 14, Folder 148, RAC.

29. See the letter from the merchants of Tierra Blanca to the director of the Department of Public Health, August 17, 1922. See also the communication from the merchants of Veracruz to the public health department, October 5, 1921, AHS.

30. An example of these appeals is the letter from workers' organizations to the President of Mexico, Veracruz, August 14, 1923, Folder 10, AHS.

31. Warren to the International Health Board, July 15, 1926, RFA, R.G. 5, Series 1.2, Box 258, Folder 3282, RAC.

32. F. F. Russell to Ms. Read, September 18, 1925, RFA, R.G. 5, Series 1.2, Box 226, Folder 2872, RAC.

33. Joseph M. Gilbert, "Revolution from Without: The Mexican Revolution in Yucatan, 1910–1940," in Edward Moseley and Edward D. Terry, eds., *Yucatan: A World Apart* (Alabama: University of Alabama Press, 1980), 145.

34. Moseley and Terry, *Yucatan.*

35. Gilbert, "Revolution from Without."

36. Moseley and Terry, *Yucatan*, 3. See also James C. Carey, *The Mexican Revolution in Yucatán, 1915–1924* (Boulder, Colo.: Westview Press, 1984).

37. John Womack, Jr., *Zapata and the Mexican Revolution* (New York: Vintage Books, 1968).

38. Salamini, *Agrarian Radicalism.*

39. Carey, *The Mexican Revolution.*

40. The petition was supported even by henequen planters. Gilbert, "Revolution from Without," 143; and Carey, *The Mexican Revolution*, 85, 62. It is interesting to note that Alvarado visited the United States on several occasions, and Carrillo Puerto spoke English fluently and maintained a romantic relationship with North American journalist Alma Reed. Joseph M. Gilbert, "The United States, Feuding Elites, and Rural Revolt in Yucatán, 1836–1915," in Daniel Nugent, ed., *Rural Revolt in Mexico and U.S. Intervention* (San Diego: Center for U.S.-Mexican Studies, University of California, San Diego, 1988), 167–97.

41. The foundation's concern about the yellow fever in Yucatan came as early as the construction of the Panama Canal. "Yellow Fever: Feasibility of Its Eradication," report sent by W. Rose to Dr. J. H. White, August 29, 1914, RFA, R.G. 5, Series 1.2, Folder 88, RAC.

42. W. Rose to Mr. Stabler, July 19, 1918, RFA. R.G. 1, Series 305, Folder 15, Box 2, RAC.

43. M. E. Connor to W. Rose, September 3, 1918, RFA, R.G. 5, Series 1.1, Box 31, Folder 530, RAC.

44. Connor, "Notes on Yellow Fever."

45. The information appears in T. C. Lyster, Diary, December 16, 1919, RFA, R.G. 12, RAC.

46. According to an RF doctor, "Bolshevism has complete sway in Yucatan. Strikes occur for any trivial reason." Ibid.

47. "The native population has . . . in the past all acquired their immunity in childhood so have no fear of the disease." T. C. Lyster, Diary, December 16, 1919, RFA, R.G. 12, RAC, 81–83. See also George C. Shattuck, *The Peninsula of Yucatan: Medical, Biological, Meteorological and Sociological Studies* (Washington D.C.: Carnegie Institution of Washington, 1933).

48. Memorandum from Dr. Rose to Dr. J. H. White, February 24, 1914, RFA, R.G. 5, Series 1.2, Box 6, Folder 86, RAC.

49. Shattuck, *The Peninsula of Yucatan.*

50. M. E. Connor, "Preliminary Reports, Yellow Fever, Merida Yucatan, Mexico," June 22, 1921, RFA, R.G. 5, Series 3, Box 147, RAC. See also Shattuck, *The Peninsula of Yucatan*, 84.

51. For a discussion of this topic, see Marcos Cueto, "Sanitation from Above: Yellow Fever and Foreign Intervention in Perú, 1919–1922," *Hispanic American Historical Review* 72 (1992): 1–22.

52. M. E. Connor to T. C. Lyster, August 16, 1921, RFA, R.G. 5, Series 1.1, Box 56, Folder 806, RAC.

53. The validity of the vaccine, and the identification of *Leptospira icteroides* as the organism producing the fever, was inaccurate and wrong. For an analysis of this controversy see Juan Guiteras, "Expedición al Africa y Estudios de Fiebre Amarilla," *Revista de Medicina y Cirugia de la Habana* 26 (1921): 95–115; and George K. Strode, ed., *Yellow Fever* (New York: McGraw-Hill, 1951).

54. The RF had previously employed the technique of using larva-eating fish in its cam-

paign in Guayaquil, but this method became more attractive in Yucatan where there were plenty of larva-eating fish. See Cueto, "Sanitation from Above."

55. Yellow Fever Report 1920–1921, March 1921, RFA, R.G. 5, Series 3, Box 147, RAC.

56. Ibid.

57. Progreso de Castro, "Departamento de Salubridad Pública de la Nación: Servicio Especial de la Fiebre Amarilla en Yucatán," August 1920, Folder 23, AHS.

58. While the cost of depositing a fish ranged from two to three cents per fish, the cost for depositing oil varied from $7.41 to $13.70 per deposit. See M. E. Connor, "Narrative Report of the Yellow Fever Control Work in Zone No. 1. Mexico," RFA, R.G. 5, Series 2, Folder 194, RAC.

59. M. E. Connor, "Yellow Fever Reports, 1922," RFA, R.G. 5, Series 3, Box 147, RAC.

60. M. E. Connor, "Paper Delivered before the Association of School Teachers at Progreso, Yucatan, Mexico" (original in Spanish), October 10, 1922, Folder 16, AHS.

61. Dr. Pedro F. Rivas to the Rockefeller Foundation, May 16, 1922, RFA, R.G. 5, Series 1.2, Box 137, Folder 1819, RAC.

62. The information appears in T. C. Lyster, Diary, December 16, 1919, RFA, R.G. 12, RAC.

63. Later, Dr. Hernandez accepted a position at the Rockefeller Institute, with the understanding that his nomination represented a "patriotic pride and a distinction to all Mexicans." Dr. Diego Hernandez to Dr. Angel Brioso Vasconcelos, February 14, 1923, Folder 32, AHS.

64. Reported in *El Universal*, a Mexico City newspaper, 1922. No specific date is given on the clipping, which is kept in the section on yellow fever, Folder 15, AHS.

65. The Mexican Army guaranteed the peaceful and stable conditions that U.S. investors were demanding: "peace, oder, and conditions conducive to work and trade." Carey, *The Mexican Revolution*, 209.

66. See the newspaper, *El Dictamen* (Veracruz), December 1922.

67. The information appears in Wickliffe Rose, Diary, December 28, 1920, RFA, R.G. 12, RAC.

68. Ibid. Some Mexican doctors working for the federal government agreed with and celebrated the foundation's decision, "General Plan for the Yellow Fever Campaign 1922," December 1921, Folder 37, AHS.

69. It is important to note that Mexico had already established a Secretariat of Public Health, but this office was far from what the Rockefeller Foundation's representatives had in mind. It was Wickliffe Rose, director of the Rockefeller Foundation, who outlined the specific characteristics that a state health department should possess.

4

Exporting American Agriculture

The Rockefeller Foundation in Mexico, 1943–1953

Deborah Fitzgerald

BOTH CRITICS AND champions of the Green Revolution agree that Western aid to developing nations has had a profound impact not only on the latter's agricultural productivity, but on their social structures as well. But few analysts recognize the extent to which these results are fundamentally based in the institutional models used in ordering agricultural development in these countries. While some have argued that the technologies exported to developing nations are inappropriate, one might extend the argument by locating the inappropriateness in the institutional structures and ideologies from which these technologies have emerged.

In recent years, as economists and policymakers have evaluated the assets and liabilities of the many development strategies that fall under the rubric of the Green Revolution, several scholarly biases have emerged. First, perhaps the most glaring problem in development studies is the concentration on Indian and Asian projects. While there are several good explanations for this, not least of which is the very real concentration of capital, commodity, and human resources there, such a focus tends to oversimplify the difficulties encountered in developing nations generally. That is, these studies seem to suggest that not only are the problems of technical adaptation comparable from one culture to another, but that the Indian or Asian experience might serve as a model for technical interventions in other countries. Second, historians have in large part refrained from entering the agricultural development territory, leaving it to economists and policymakers whose sense of historical context is often somewhat undeveloped. Thus, we are led to believe that the Green Revolution "began" in the 1960s, despite the fact that programs of technical assistance were introduced in Latin America in the 1940s and 1950s.[1] This ahistorical tendency is perhaps best exemplified by what one prominent economist termed technology that is "value neutral as to scale."[2] While it is debatable whether or not technology is value neutral under any circumstances, within the actual context of agricultural introductions, a technology (or scientific expertise, or institutional structure) designed for one particular context and introduced wholesale into a different context can in no way be considered value neutral for scale or any other attribute. The technical design and specifications of such technologies tend to reflect the social and economic conditions of the regional market for which the technology was developed, and thus incorporate assumptions regarding natural resource availability, community needs and interests, local and national agricultural priorities, and so forth.

In this chapter I shall attempt to provide a historical context for such issues by examining in detail one of the first programs in technical assistance, that of the Rockefeller Foundation (RF) program in Mexico. Begun in 1943 and ending in 1961, this program provides a useful historical device for examining the notion of agricultural development in several ways. First, it was one of two United States–based technical assistance programs exported to Latin America in 1943, the first year in which foreign agricultural problems were systematically addressed by means of sophisticated, in situ, agricultural technology and expertise.[3] As one of the first such programs, and as one considered a success by the late 1940s by most analysts, it can arguably be considered a model for later aid programs. Second, while the Mexican program represented the RF's first coherent experience with agricultural development, it led directly to their patronage of similar programs throughout South America and ultimately to the establishment of the International Maize and Wheat Improvement Center (CIMMYT), a world-renowned agricultural research program, now in itself a model for technical assistance.[4] Although CIMMYT has tended to obscure the more modest Mexican program, its focus and structure were based deliberately on the most successful aspects of the Mexican program: research and training.

Yet not all areas of the Rockefeller program were successful, and here I shall argue that the pattern of success and failure in Mexico resulted from the RF's reliance on a peculiarly American model of agricultural development. That is, in attempting to strengthen Mexican agriculture, the RF used not only American experts, seeds, and irrigation systems, but a set of assumptions and expectations about how agricultural systems respond to such "inputs," and how institutions and individuals behave when assistance is offered. The American system, characterized by relatively large farming tracts operated by middle-class families, was in the 1940s capital- rather than labor-intensive, market- rather than subsistence-oriented, and entrenched in a network of commercial and public institutions designed to make farming more productive, efficient, and businesslike. Mexican agriculture was, by and large, quite different; yet while most Mexican farmers were subsistence-oriented, labor-intensive, and unconnected to any larger system of agricultural advice and expertise, a small proportion were in many respects similar to American farmers both in attitudes and circumstances. The difference between these two groups underlines my argument: as a rule, the RF program succeeded insofar as Mexican farmers and agricultural professionals shared with their American counterparts the same assumptions about and expectations for agricultural progress, and failed when such a parallel group did not exist. This is not to suggest that the RF self-consciously and deliberately attempted to mold the Mexican agricultural structure in the image of the U.S.A.; rather, their attempts to bring the most successful elements of American agriculture to Mexico were informed by their own agricultural experience, and thus their achievements reflect the similarity between the American and Mexican agricultural situations.

The Historical Background

For the Rockefeller Foundation, funding agriculture represented a new but not unlikely direction in patronage. From 1902 until the mid-1920s, RF programs often resembled relief efforts. For the most part they were a response to some specific need for food, medical care, public health, or education. Rather than establish long-term and complex programs, they tended to provide stopgap emergency aid to people and institutions that could eventually rely on federal assistance. From the late 1920s through the 1940s, the RF concentrated its philanthropic efforts on scientific research that could be applied to social and practical problems, especially in education and public health. The RF was particularly interested in those disciplines not already supported by federal and industrial funds; committing itself to research areas that were in a nascent and therefore malleable state, the RF hoped to "rationalize the social order" by encouraging the production of useful knowledge.[5]

The RF's introduction to the potential of agricultural science for promoting both rural betterment and a viable farm economy was their cooperation with a USDA program in 1906. The 1906 program, under the direction of Seaman Knapp, was a vigorous crusade against the boll weevil in the southern U.S.A. that emphasized the importance of scientific farming practices.

Knapp's strategy for improving agricultural practice was farm demonstration, whereby strategically located farmers were persuaded to adopt a package of practices that generally included improved seed and increased cultivation, fertilization, and mechanization.[6] In time Knapp also directed his efforts at farm children by setting up Boys' Corn Clubs and Girls' Canning Clubs, the aim being twofold: first, he hoped that participation in these programs would impress upon farm youth the benefits of scientific farming, a lesson they would remember as adult farmers; and second, the program was an attempt to convert the more reluctant and cynical farmers to the Knapp system. He reasoned that no farmer would want to harvest a yield poorer than that of his son.

The RF's involvement with the USDA program from 1906 to 1919 was an extension of their General Education Board program in the South which focused on hookworm eradication.[7] Since the USDA was primarily interested in combating the boll weevil and only secondarily in farm demonstrations, they would not finance the Knapp program in non-weevil-infested states. The role of the RF was to provide funds necessary to promote farm demonstrations in other southern states. Although their role was almost completely that of patron, they were as proud of their involvement as if they had conceived the idea themselves. Underlying their cooperation with the program, and in part explaining it, were three assumptions. First, they believed that the most pressing need in the rural South was economic. Without money teachers were inadequately prepared, education poor, and farmers so impoverished that field preparation was ignored and yields remained low. Second, they felt that involvement with Knapp's work would be finite, that dramatic results would be achieved in a few years, at which time the local, state, and county authorities, impressed with the newly demonstrated potential for southern farming, would accept the full burden of the program. And third, they decided that the thrust of the program should be aimed not at future farmers, the research community, or the rural merchants, but at those who were actively engaged in farming. The effort to revitalize southern farming, to make it an economically viable lifestyle and occupation, was thus made in a very direct manner. Once the improvements were made available, the plan was for horizontal rather than vertical expansion: from farmer to farmer through emulation, rather than from expert to farmer through education.[8] This demonstration work in the South, which, in a circuitous way, inspired the Smith-Lever Extension Act ten years later, was an important component of the land-grant university system, which itself constituted the institutional structure and ideological basis of American agriculture.[9]

The land-grant system as recognized today resulted from a long series of negotiations between federal, state, and local governments, as well as farm groups, agribusiness, and agricultural professionals.[10] The Morrill Land-Grant Act of 1861, by which the federal government provided the means for each state to develop an agricultural college, was followed by four other acts that allowed the colleges to expand and diversify. The Hatch Act (1887) provided funding for experiment stations, the Adams Act (1906) provided funding for research, and the Smith-Lever Act (1914) provided funding for the extension of scientific knowledge and practice to farmers. Unlike the basic liberal arts colleges, the land-grant college offered extensive vocational training in agriculture and home economics that early planners thought reflected the needs and concerns of farm families. The educational curriculum thus reinforced the agrarian, home-centred values imputed to farm families, and extended the scope to these cores of knowledge by transforming them from commonsense practices to scientifically grounded spheres of expertise. The research component grew directly from this educational agenda, both in the college and the agricultural experiment station.[11] In both contexts, research was ostensibly geared to improving rural conditions through applying sophisticated scientific analyses to farm and home products, and providing technological innovations that would reduce the labor required in producing such products. The extension service, then, brought these programs full circle. By taking research results to farmers in the form of practical and innovative recommendations, the agricultural professionals were determining the value and pertinence of their research and receiving critical feedback from the farm community. Their frequent presence in the community also encouraged farm youth to seek their fortunes in scientific, rather than traditional, agricultural practice. While this model varied somewhat from one state to another, especially insofar as farm commodities differed regionally, it was remarkably predictable overall.

Yet despite the rhetoric of land-grant administrators, who suggested that the system and the scientific knowledge it produced would democratize agriculture, the land-grant system was dominated by commercially oriented scientists and farmers. These participants were less concerned with educating ordinary farmers to become more efficient than with fashioning sophisticated scientific and technological agenda that favored farmers who could afford to experiment and expand. Eugene Davenport, Dean of the Agricultural College at the University of Illinois (1895–1922), held a fairly typical view on the role of the colleges:

> I do not agree with the proposition that the college . . . should . . . take the message to every individual, on the principle of letting no guilty man escape. There will always be lost souls in farming . . . and there will be men not worth saving; for this is public business and not charity.

For Davenport and others, extension was designed for those farmers who had the knowledge and wit to ask for assistance, and the financial capacity to experiment with agricultural innovations dreamed up by the research staff.[12]

The Mexican Program

The Personnel

The first signs of RF interest in Mexico appeared in early 1933 when John A. Ferrell of the RF's International Health Board (IHB) discussed the prospect with Mexican officials in the public health sector.[13] Again, in 1941, the issue was raised with the IHB chief in Mexico, George C. Payne, and later in the year with U.S. Vice-President Henry Wallace. All agreed that the primary need in Mexico was for a program covering public health, education, and agriculture. But when the RF staff met later in 1941 to decide whether or not to become involved with Mexico, plans for both education and public health were abandoned. Formal education was considered too controversial, and a public health program was already underway in the IHB.[14] The program therefore became a purely agricultural one.

The agricultural professionals the RF chose to design and implement the Mexican program were fully entrenched in both the land-grant system and the philosophy of expansion it represented. The first group appointed, in July 1941, was the Survey Commission, whose task was to evaluate the Mexican agricultural situation firsthand. This group continued to act as the Agricultural Advisory Committee for the Mexican program after its return, and was responsible for policy development in the first decade of the program. The three-man team included Elvin C. Stakman, a plant pathologist from the University of Minnesota; Richard Bradfield, an agronomist and soils expert from Cornell; and Paul C. Manglesdorf, a plant geneticist from Texas A&M. All three advisers came from the land-grant system, both as students and professionals working in agricultural experiment stations, and all held positions within the USDA more generally.[15] Stakman, without question the most assertive of the three, had acted as a USDA agent to Latin America since 1918, traveling each year to assess the development of wheat stem rust. In addition he acted as scientific adviser to the Firestone Plantations Company in Liberia in 1930.[16] Although Stakman was the RF's first choice for heading the Mexican program in 1943, he declined and instead recommended his former student, J. George Harrar, who accepted the position. Harrar, a plant pathologist at the University of Washington in Pullman, then chose his scientific staff: Edwin J. Wellhausen, corn (i.e., maize) breeder; Norman E. Borlaug, plant pathologist; William E. Colwell, soil scientist; John J. McKelvey, economic entomologist; and Lewis A. Roberts, corn (maize) breeder.[17] Considered as a group, the advisers and scientific staff embodied the land-grant university experience. Seven were born in the rural Midwest, one in rural Texas, and one in rural New York. All received at least part of their education in the land-grant colleges, and five went on to work in agricultural experiment stations. Several were educated by others in the group: Stakman taught Harrar and Borlaug, Manglesdorf taught Roberts, and Harrar taught McKelvey. While only Wellhausen had been involved with the RF before the Mexican program, all but Colwell remained in one capacity or another; Harrar went on to become foundation president in 1961.

It is important to note here that although the RF did not discuss their patterning of the Mexican program on the land-grant model, such an explicit account is irrelevant for our purposes. In the 1940s, sophisticated agriculture and land-grant strategies were essentially one and the same thing in the U.S.A., and there was ultimately only one plausible model to which these scientists could refer. Thus, the advisers and staff quite naturally drew on their experience when ordering their new experiences and priorities in Mexico, in effect building a Mexican land-grant network as a reflection of their own. They identified the preexisting structures in Mexico as parallel to those in the American context: in place of the USDA they had the Mexican Ministry of Agriculture; in place of the land-grant college they had the agricultural colleges of Mexico; in place of the experiment stations they established their own Office of Special Studies; and, of course, the Mexican farmer replaced the American.

Early Problems

Following a two-month tour of Mexico, in 1941, the RF's Survey Commission prepared a report identifying the major problems of Mexican agriculture and recommending solutions. They felt it was important to maintain a four-man commission in Mexico consisting of an agronomist, a plant breeder, a plant pathologist and entomologist, and an animal husbandryman. This commission, in close cooperation with the Mexican Ministry of Agriculture, would conduct experiments and research, establish regulatory practices, and teach farmers through extension and adult education. More specifically, the Survey Commission identified four primary problems whose solution was critical to the success of the program: the improvement of soil management and tilling practices; the introduction, selection, and development of superior varieties of grains and legumes; the control of pests and diseases on a national scale; and the introduction of better breeds of farm animals, chiefly cattle. As the program developed, regulatory practices were not formally enacted, adult education as such was never attempted, and animal production was not systematically addressed. The focus of the program instead became research and advanced training.[18]

When the program started in 1943, with Harrar's team transforming the commission into the Office of Special Studies (OSS), the Mexicans had mixed feelings about the role of the RF in their country. By all accounts, officials at the Mexican Ministry of Agriculture were pleased to have not only a revitalization of agricultural science, but also the input of RF dollars. According to the contract between the RF and the ministry, each group would contribute roughly equal shares to the program, with the RF providing trained personnel and equipment and the ministry donating land and labor.[19] Agricultural scientists, on the other hand, were not so enthusiastic. According to Harrar, their ambivalence was due to both national pride and the nonexperimental character of Mexican science. Harrar's opinion was that Mexican agricultural education relied almost entirely on textbooks, with little or no fieldwork required for a college degree. The Mexicans seemed to feel that agriculture was a science best learned in the laboratory and classroom rather than the field, and a social stigma was apparent toward fieldwork generally. Agriculturalists were sceptical that anything could come of field experimentation, feeling that the Americans were naive in resorting to such "primitive" practices. In addition, many agriculturalists were ill-disposed toward foreigners who set up shop in their country, an act that seemed to suggest that Mexican scientists did not know what they were doing.[20]

The OSS group found still others in the government and academe harboring unrealistic expectations of American science. The Mexicans had heard of the wonders of American agriculture—that yields were huge and quality superior—and such science had come to seem almost mystical. Some wondered if the Americans had a special knowledge that produced miracles. Certainly the hyperbole that surrounded the diffusion of hybrid corn during this period in the U.S.A. would lead easily to such misapprehensions.[21]

Partly because of these suspicions and expectations, the OSS for a time had trouble identifying itself to the Mexican farmers, especially those whom they addressed on field days and at farm demonstrations. Harrar felt that they could not say they were an arm of the government, for no one trusted them; farmers were not familiar with the notion of philanthropy, and the RF team did not want to be mistaken for missionaries. As Harrar put it, the OSS group was often thought by the Mexicans to be "not too bright." In time they overcame this problem by employing Mexican personnel wherever possible.[22]

Maize (Corn)

Once installed in the OSS, Harrar and his staff began tackling the many agricultural agendas set out in their agreement with the Mexican government, six of which were significant in developing later agendas: corn breeding, wheat breeding, soil improvement, plant protection, extension, and education. When Harrar arrived in 1943, there were two programs in seed corn improvement already underway. One was operated by Edmundo Taboada of the ministry, and the other by Eduardo Limon at the Leon Experiment Station. Both programs were begun in 1941 at the suggestion of Henry Wallace, but neither had achieved much success.[23] The lack of financial and material resources, as well as the attempted use of American seed corn, seriously hampered the progress of these programs. One thing they demonstrated, however, was that American varieties were not suitably adapted to Mexican climate and soil, information that was to play a major role in the development of OSS breeding experiments.

The OSS corn research program developed three approaches in an attempt to improve the raw material resources of the Mexican corn grower. The most rudimentary, low-technology strategy was also the most critical to the success of the program overall. It consisted of testing hundreds of native corn samples for vigor, growing period, climatic adaptability, yield, and other cultural factors. Superior strains were selected and their seeds increased to create a sufficient quantity of corn for distribution and planting. This effort was important because it not only supplied farmers with improved seed very quickly, but also isolated those native strains that would contribute most to hybrids and synthetics. Synthetics were developed by crossing inbreds or single crosses with superior open-pollinated varieties. They were thus intermediate between ordinary, heterozygous open-pollinated corn and the more homozygous hybrids. The virtues of synthetics were that they were more adaptable than hybrids to different climatic zones and, because of their broader genetic diversity, their seed could be used in planting for several years without significantly reducing yield and vigor. Thus synthetics were particularly well suited to the Mexican context where corn farmers were less affluent than their American counterparts, and the climatic regions more diverse. The third corn strategy was, of course, hybrid corn; by 1948 ten hybrids were available to Mexican farmers.[24]

In 1947 Ricardo Acosta, a commercial farmer and graduate of Chapingo, established an agency called the Corn Commission which attempted to get superior strains from the OSS, give these to private farmers for seed increase, and then collect and distribute the seeds. Although not an official government office, and not associated with the ministry, the commission had the support of then-President Miguel Alemán. Several of the RF advisers, notably Harrar and Stakman, were apprehensive about giving seed to the commission, fearing that it might overcharge farmers for the seed, or distribute seed with no concern for the fact that particular seed had been bred for particular climatic regions. Nevertheless, despite their worries, the OSS joined the partnership and turned the seed over to the commission. Although their specific fears proved to be unfounded, their general apprehension was not, for a fierce competition developed between the commission and the ministry. Don Nazario of the ministry set up a rival increase and distribution center called the National Commission for the Increase and Distribution of Improved Seed. President Alemán changed his official support from Acosta to Nazario, ordering the OSS to give their seed to Nazario rather than Acosta. Exasperated by this political jockeying, the OSS finally agreed to give an equal share to both groups.[25] But their problems were not over. Although Nazario's program declined in importance, Acosta too had his share of trouble. Within a few years his director died and the ministry seized the opportunity by placing one of its men in his post. The infighting that developed between him and Acosta took its toll on the efficiency of the program, and the commission never did live up to its earlier promise.[26]

Evaluating the assets and liabilities of the corn program is difficult. Although 1948 marked the first year in which Mexico did not import corn, due in large part to the use of improved seed and a particularly favorable growing season, this may be a misleading signpost.[27] On the international scene, the Mexican government did not have an advantageous price support system that might have enabled farmers to compete in the export market.[28] On the national scene, most farmers simply did not adopt improved seed; even by 1963 less than 12 percent of Mexico's corn acreage was planted with hybrids, or 36 percent if synthetics and selected varieties are included.[29] Despite sophisticated research strategies, Mexican farmers were not incorporated into the network. Most of the corn in Mexico was grown on subsistence-level farms averaging three hectares. Farms of this size were not in a position to adopt expensive practices such as hybrid seed, fertilizer, and the irrigation system often necessary to make the hybrid investment worthwhile. In addition, these small corn farmers were widely scattered throughout the country; for them to adopt hybrid corn would have necessitated an elaborate extension network and a wide range of improved strains suitable to many particular climatic zones. The corn program did not in fact have a sophisticated extension network; those farmers who adopted OSS practices and recommendations were not representative of corn growers in general. When considered in terms of the land-grant model, it seems clear that the OSS experimental station aspect worked well in producing research geared to particular areas but failed to extend its expertise when faced with a farm community very different from its American counterpart. According to one RF participant, the RF staff did not build up a comprehensive research and extension program in corn, and thus were not successful in promoting adoption.[30]

Wheat

The situation with wheat, however, was quite different. When Harrar and Stakman toured the wheat-producing areas of Mexico in early 1943, they found the yields uniformly discouraging. The reasons for this, they felt, were several: the use of poor varieties, depleted soils in the north and central regions, the presence of stem rust in fertile areas, and the lack of plant nutrition. As with corn, Limon and Taboada were already at work on the wheat-rust problem. In 1942 the ministry began a program of testing wheat varieties for rust resistance, but a declining interest produced only very limited results.[31]

In 1943 Harrar began a program designed to create wheat varieties resistant to stem rust.[32] His group collected both wheat seed and samples of stem rust, and in the fall planted 700 varieties, both Mexican and American, inoculating them all with the rust. The following June they replanted the 500 survivors. From this they learned that while Mexican varieties ripened early (which makes multiple cropping possible), they were not resistant to stem rust. For the American varieties the inverse was true: although they did not ripen early, they were resistant. To solve this problem Norman Borlaug joined the program in 1945. Since neither pure Mexican nor pure American strains were suitably efficient, Borlaug's aim was to cross them in an attempt to combine early ripening with rust resistance in a hybrid. But, for the present, he felt that the strains which survived the fall planting would be superior to strains in current use, and these improved strains were distributed to farmers in 1945. By the fall of 1948 five of the new lines were converted into varieties and distributed to farmers; their primary virtue was a resistance to stem rust. In 1945 Borlaug also began crossing superior varieties and by 1949 had tested, increased, and distributed four new hybrids.

But as with the corn program, wheat research suffered difficulties external to the breeding operation. As early as 1944 competitive tensions existed between Limon and Taboada. Caught in the crossfire, the OSS tried, albeit without much success, to nurture a three-way cooperation. And in 1949 millers discriminated against the improved varieties because of their allegedly poor milling quality. Tests conducted in the U.S.A. disputed this claim, and pointed to building a wheat-testing laboratory in Mexico to alleviate suspicions.[33] On the other hand, the improved wheat was popular with farmers and in some areas a black market in wheat emerged in 1948. In 1947 the ministry took over the increase and distribution program, and in 1950 the Wheat Commission was absorbed into the Corn Commission.[34] Despite problems with stem rust, millers, and the politics of bureaucracies, 90 percent of the Mexican wheat acreage was planted with improved varieties and hybrids by 1957.[35]

This high rate of adoption, and the correspondingly low rate for corn, has an explanation. Unlike Mexican corn farmers, wheat farmers bore a resemblance to American farmers in several key respects. They were commercial rather than subsistence farmers, each farming an average of seventeen hectares of land.[36] This increased productive capacity allowed them to take advantage of credit opportunities, invest in improved seed, fertilizer, and most importantly, irrigation systems which together comprised the package of agricultural practices recommended by the OSS.[37] In addition, the large growers were eager to experiment with new farming practices, and made an effort to keep abreast of latest developments. They did not rely on extension efforts to advertise the new ideas, but in fact sought them out.[38] The Mexican land-grant model was successful, in short, because the wheat farmers were in many ways similar to American farmers. In terms of their affluence, access to resources, and interest in experimental practice, they resembled American hybrid corn adopters rather than other Mexican farmers, and this similarity created an effective "fit" between the wheat farmers and the OSS. Interestingly, both American hybrid corn and Mexican wheat were adopted at a similar pace: both innovations were virtually completely adopted ten years after their introduction.[39]

Other Areas

OSS programs in other practical agricultural areas also reflected the difficulties of the land-grant model in Mexico and the ambivalent focus of the program itself. While the OSS could identify large-scale agricultural problems, they were unable to generate broad-based solutions to these problems within their framework. Most of the problems were so general that practical solutions to them were implausible within the framework of an American model that depended upon the enthusiasm of commercial farmers. Without an extension network capable of imposing new practices on subsistence farmers, the impact of these programs remained negligible. For example, in 1944 agronomist William Colwell joined the OSS and for the first several years concentrated his efforts in three areas of soil improvement: soil management, soil analysis, and forage crop research. He found that 90 percent of Mexican soils were nitrogen deficient because of poor soil-management practice. Many farmers planted the same nitrogen-depleting crop on the same soil for many years running. Others simply moved regularly; they would clear a field, farm it a few years, and then abandon it altogether. The ideal situation, the OSS felt, was to combine rotation of crops with chemical fertilizer, but both practices were designed for the commercial farmer. Rotation required having enough land to allow a field to lie fallow; fertilizer was expensive and, in most farming regions, its effective use required an irrigation system. Even those farmers who could afford fertilizer had difficulty procuring it. Most of it was imported from the U.S.A. by way of the USDA, and the allotments were reportedly never enough to satisfy the demand. At the request of the Mexican government, the RF agreed to intervene with the American government in a request for larger allotments of fertilizer.[40] But the problem continued. A soil improvement program was established by the ministry in 1949, but they had trouble funding it and tangible results were negligible for our period.[41] While nitrogen fertilizer use increased over twenty-five times from 1949 to 1966 overall in Mexico, only 15 percent of the area harvested received treatment.[42] This adoption rate corresponds roughly to the commercial wheat-farming population which, like American farmers, was financially able to adopt such recommended practices.

A program for crop protection was in operation in Mexico in 1943, but again it was less than effective.[43] By 1941 the ministry had established thirty-two regional offices as part of the Plant Sanitation Department, but a combination of factors prevented these offices from successfully combating pests and diseases when problems arose. For example, a banana disease entered the country in 1941 but, even though the offices were equipped with chemicals to control the disease, they were not used because of administrative problems and unfamiliarity with both the field methods required and the equipment at their disposal. Another problem in 1941 was an outbreak of wheat and barley smut. Here the strategy of seed disinfection was sound, but the equipment was so antiquated that the treatment was ineffective. In addition a great deal of disinfectant was used on corn seeds which neither had the disease nor were susceptible to the treatment.

For their part, the OSS responded to pests and diseases such as wheat-rust with an essentially "top down" approach: lacking an effective extension network, they concentrated instead on training plant protectionists and breeding resistant plant strains. There is no evidence that they attempted to educate farmers in rudimentary crop protection, and during this period only commercial farmers availed themselves of improved strains. Perhaps because major regional pest and disease outbreaks in the U.S.A. are handled by a coalition of local, state, and federal agencies in cooperation with farmers, the OSS seemed to feel that large grassroots campaigns were not part of their agenda. Lacking both a cohesive and responsive farming population and a sophisticated federal apparatus, the OSS chose to make the Mexican agricultural system resemble its American counterpart by bypassing the vexing problems of rural poverty, lack of education, and discontinuous government policies. By 1963, the ranks of degree-holding Mexican plant protectionists rose from nought to fifty, in large measure because of the RF educational program, and into their laps these difficulties fell.

Extension

One of the most severe obstacles in all the programs discussed so far was the lack of an effective extension mechanism; the problems of applying the

American land-grant model to Mexican agriculture were nowhere more apparent. While some writers have suggested that the RF program lacked an extension component, it in fact did sporadically attempt to provide extension services.[44] In 1947 the OSS held farm demonstrations across the country designed to illustrate the virtues of improved seed and farm practice. According to Harrar, attendance at these demonstrations was very poor until Mexicans themselves led the proceedings and a free dinner was thrown in for those attending.[45] By 1945 the OSS recognized the need for an extension bulletin similar to that issued by the U.S. experiment stations. In the following year the ministry agreed to print and distribute such bulletins, if the OSS would write them. By 1947 they had published two bulletins, each of which was prepared in two versions: one, written at the popular level in Spanish, was distributed to farmers; another, written in a more technical language, was distributed to agricultural scientists in Mexico and the U.S.A. By 1949 eight such bulletins had been printed.[46] But ultimately it is difficult to evaluate this effort. Certainly, the bulletin was useful to agricultural scientists because it not only told them what current research was producing, but also may have played a role in shaping other Latin American agricultural programs. As for the Mexican farmer, it was probably quite helpful to the commercial farmers who were enthusiastic about increasing their returns; but for the small farmer it presumed an ability to read, a skill most did not have. In 1948 the OSS brought Mortimer F. Barrus into the program in an effort to galvanize extension efforts. Barrus traveled around the country establishing experimental plots, giving slide shows and lectures, and performing field demonstrations—in short, acting as a county agent. But by the following year poor health forced him to retire, and his program slid into neglect.[47]

Extension work was a continuing problem. A government extension office was in operation when the OSS arrived in Mexico, but those in charge had neither the resources nor the practical experience necessary to operate extension properly. The OSS perceived this as a Mexican rather than an OSS problem: they felt that extension was the most straightforward task facing the Mexicans, and that the OSS should not attempt to do anything for them that they could do equally well for themselves.[48] The OSS trained many Mexicans who were then absorbed into the ministry's extension program, but the program always lacked the resources needed to equip and maintain its activities. Year after year the advisory committee urged the ministry to establish a more effective program in extension, and year after year the money was not made available.[49]

The RF's emphasis on self-help and federal subsidies for extension also echoed difficulties with extension in the U.S.A. The Smith-Lever Act of 1914 provided experiment stations with federal funds to develop outreach programs; until then extension work was often a last priority, with experiment station scientists reluctant to relinquish research time.[50] But while in the U.S.A. the problem was too much farmer demand on researchers, in Mexico it was too little. Typically, the successful farmer relied not on extension efforts but on his own entrepreneurial initiative, while small farmers were often unaware of services available.[51] Thus, in the U.S.A., policymakers were inundated with farmers' demands for advice and, consequently, experiment station scientists' complaints that their research time was consumed by public service responsibilities. In Mexico, farmer demand was virtually nil and researchers were rarely called upon to do general public service. Neither the Mexican government nor the RF was sufficiently inspired to generate an effective extension service. In addition, the American extension workers often required farm groups or rural townships to request a farm demonstration before one was provided, or a particular circular or bulletin before one was sent. This attitude, which helped produce a land-grant system that provided a resource for those who asked, and a mystery for those who did not know what to ask for, characterized the programs in Mexico as well as in the U.S.A. While this logic had a certain plausibility in the U.S.A., in Mexico it was wildly inappropriate, and led to the widespread belief that there was no extension program at all.[52]

Education and Training

The RF program in agricultural education also reflected the contextual constraints of the land-grant model. The failure of the elementary education program and success of the college program further strengthen the view that OSS programs were successful insofar as the Mexican target group resembled its American analogue. In 1943, A. R. Mann toured Mexico and concluded that rural elementary schools needed assistance. The main problems were a lack of teachers and the number of different dialects that made acquiring appropriate teachers a problem.[53] In 1946, the OSS made efforts to provide farm demonstrations at elementary schools, but little came of it.[54] Warren Weaver, director of the RF Division of Natural Sciences, identified the need for a program that would employ public health, agriculture, and elementary education, but was also aware of the problems of such an approach. Primarily, he felt,

> elementary education is hot politically. There is also the difficulty,
> even in quite non-political areas, of getting into conflict with intense
> local pride and culture patterns which we may neither fully
> understand nor appreciate.[55]

Weaver was fairly sensitive to the ideological ramifications of such a program, and the RF in general avoided entering that arena. But with Henry Miller,

Weaver came up with an alternative to direct elementary education, that is, vocational agricultural training at the elementary level. Weaver was not very specific about how this would work, but one can surmise that it would have
included farm demonstration, the teaching of basic agricultural skills such as
cultivating, planting, and harvesting, and perhaps even domestic skills associated with both public health and home economics.[56] Such a program had already been developed in the U.S.A. through the extension service, and was
implemented by the RF itself in their southern work. Boys' Corn Clubs, Girls'
Canning Clubs, and the 4-H movement were essentially designed to include
farm youth in the extension campaign for scientific agriculture. In Mexico this
plan was ultimately rejected, as were all rural elementary and adult education
programs.

The most successful adaptation of the land-grant model was in advanced
education, an area where again the similarities between the U.S.A. and Mexico
outweighed the differences and where, in fact, the land-grant model was employed directly by training Mexican students within the American system.
There were two sorts of advanced training programs sponsored by the OSS:
one for students and one for professionals.[57] Fellows, as established Mexican
scientists were called, were eligible to receive advanced agricultural education
in the U.S.A. In this way existing agricultural educators could benefit by obtaining the advanced training not available in Mexican colleges when they were
students. But for the future of Mexican agriculture it was the students who
formed the core of the program, acting as a link between the OSS, the agricultural colleges, and the ministry.[58] At first the arrangement was haphazard.
When the OSS needed fieldhands, students were hired from the colleges. If
they showed promise the OSS kept them on in an apprenticeship arrangement,
each student assigned to a project leader. But not all were students. Some were
found nearly by accident, in the fields or on the street—there was no pattern.
According to Harrar, it was difficult to find workers both because of the
Mexicans' distrust and suspicions of the OSS and because fieldwork was held
in low regard. As a consequence, the OSS generally offered higher wages than
other potential employers.[59]

Eventually, a more systematic arrangement was worked out with the National College of Agriculture at Chapingo and the ministry. A state-run institution, Chapingo gave full fellowships to all who attended the seven-year program in agriculture. On graduating the student was awarded the "ingeniero
agrónomo" degree, and often such a student would then be commissioned to
the ministry. The primary problem with this program was that the student was
neither fish nor fowl; he had virtually no practical experience upon graduation,
nor did he generally pursue research. In addition, Chapingo offered no graduate training at all.[60] For such students the OSS functioned as a sort of graduate

school of agriculture. By assisting in the field and laboratory the student gained not only practical experience, but learned how a research project was conceived and carried through. For the promising student, one of two things happened after his tenure with the OSS. In the regular scheme of things, the student would be sent back to the ministry where he would assume administrative and research responsibilities, and from where he might be sent to one of the specialized regional bureaus. If he appeared research-oriented he might be sent to the U.S.A. for further education in the graduate division of a land-grant college. Acquiring a master's or doctoral degree was not necessary—after a stay of one or two years some received a degree, but more did not. It seems to have depended largely on the facility with which they learned the English language.[61] After this the students returned to Mexico where they were commissioned to the ministry. Some went to work for the Corn Commission, some worked on the government's dam and irrigation project, some went into forestry or seed increase and production, some became ministry officials or college professors, and a few returned to the OSS as researchers and extension agents. By 1954, over 200 students had passed through the OSS program, and the majority went from there into some other aspect of Mexican agriculture.[62]

Administratively, the OSS both selected the students to receive training in the U.S.A. and paid for their tuition and expenses. The ministry's end of the bargain was to forfeit the students for a period of four or five years following graduation, a time in which the ministry would normally employ them as commissioned personnel. The ministry was also expected to arrange for the student's employment after training was completed. If the program faltered anywhere it was here, and by 1953 this coordination was still something of a problem.[63] By that time, however, another problem was developing as well. In part due to the training received by Mexicans in the OSS, by 1950 the office found itself competing with government and commerical operations to employ their own graduates. Often, after a student had completed training with the OSS and was waiting to be sent abroad for advanced work, he would be enlisted by the government to work on what it termed emergency situations. For this reason the OSS had to postpone sending students to the U.S. for a few years in the mid-1950s.[64]

Once the training program was established, the OSS discussed the possibility of inducing the ministry and Chapingo to grant a graduate degree to those students trained by the OSS. This was first suggested by Stakman in 1946 and again by Harrar in 1950, but there is no evidence that it ever happened. The obstacle here appeared to be the attitude of professional and administrative staff at Chapingo, who felt threatened and resentful of RF meddling. It was in this context especially that some within the college accused their colleagues of allowing American goals to take precedence over Mexican interests. It appears that this program was abandoned precisely because of such pressures.[65]

OSS and the Colleges

Apart from the graduate program idea, however, relations between the OSS and the agricultural colleges appeared friendly. Despite the fact that the RF preferred to fund "men not buildings" they built and equipped a greenhouse at Chapingo in 1946 that was designed for OSS research in entomology, plant pathology, and plant breeding.[66] And in 1945 the OSS staff was invited to hold weekly Saturday seminars for faculty and students at Chapingo.

Other colleges also benefited. Although the ministry expected the RF to cooperate only with Chapingo, they also aided two other colleges in the early 1950s: the Escuela Superior de Agricultura "Antonio Narro" Universidad de Coahuila in Saltillo, Chahuila, and the Escuela de Agricultura y Ganaderia Instituto Technológico y des Estudios Superiores de Monterrey in Monterrey, Nuevo León. The school at Saltillo was established with a bequest from Don Antonio Narro Rodriguez, who left his land and fortune for that purpose. Although it began as a private institution it was subsumed by the state of Coahuila in 1938. In 1952 the school began requesting assistance from the OSS in the form of money for equipment, a library, and faculty training.[67] The OSS did give them substantial aid, but in early 1953 it looked as though such aid would have to be tempered—the officials at Saltillo, and in particular the Director Lorenzo Martinez, was said to be "the most effective asker . . . ever encountered." Harrar felt that, as with all philanthropy, there was a danger in too much generosity:

> If we condition him to too much assistance, particularly that which has to do with salaries, he will not make the progress which he should make in getting necessary salary budget from the state and federal sources.[68]

The school at Monterrey was in a rather different situation. It was unique in that it was essentially the result of private enterprise, the entire school supported by "progressive industrialists." Leonel H. G. Robles, who had been an OSS student and sent to Minnesota as a fellow in 1944, was in charge of the agricultural work there. In fact, by 1952, the entire agricultural staff of eight had trained with the OSS. One of the first things Robles did when he arrived at Monterrey was to organize a "Society of Patrons" composed of industrialists, private individuals, and businesses, which helped support the agriculture curriculum. Because of this demonstration of self-help, the OSS was unusually generous to Monterey, spending $600,000 on improving conditions there in 1954. The OSS also helped in planning the curriculum.[69]

But it was not only the Mexican students and scholars who benefited from the OSS program. From 1947 to 1954, fourteen scholars from other Latin American countries came to the OSS for training, including five from the Point IV program. The OSS was also in charge of training scientists in the Central American Institute of Nutrition, an organization sponsored by the Kellogg Foundation, the Pan American Sanitary Bureau, and the Department of Nutrition at MIT. One aim of this program was to train a scientist in each of three countries to oversee research and food-crop development for the country's nutritional needs. Although the RF was not formally involved with this collaboration and did not furnish any funding, participants agreed that they were most capable of training Latin American agriculturalists.[70]

International Expansion

By 1947, an organizational model had emerged in the OSS that seemed stable enough to be copied, yet flexible enough to adapt to new environs. Despite the problems they had encountered, the advisory committee decided that the time was ripe for expansion into other countries.[71] The first offshoot of the Mexican program began in 1950 in Colombia, where the RF had given previous grants for buildings and equipment to the National University.[72] Following an initial request for aid in 1948, Lewis Roberts (director and corn breeder) and Joseph Rupert (wheat breeder) went to Colombia in 1950 to establish an agricultural research center.[73] The RF's expansion into other Latin American countries represented a distillation of their Mexican efforts and a codification of elements would form the RF's international research agenda. Two such elements—the training of agriculturalists from other countries in the OSS and coordination of international agricultural conferences—grew out of the Mexican program in the 1940s, and were critical precedents for later programs such as CIMMYT.[74]

Conclusions

The RF program in Mexico bore more than a passing resemblance to the American land-grant university system; indeed, as I have suggested, it is only by identifying its structure as "land-grant inspired" that we can coherently address the strengths and weaknesses of the program. As Delbert Myren has shown, the programs for developing and extending hybrid corn and wheat in Mexico were remarkably different from each other largely because wheat farmers constituted a small, sophisticated, geographically focused, and commercially oriented group, while corn farmers were a large, geographically diverse group restricted to subsistence farming. I would extend his analysis by locating the success of the wheat program in the cultural similarity between Mexican wheat farmers and American farmers generally, and the failure of the corn program in the dissimilarity between Mexican corn farmers and American farmers. Because the OSS employed an organizational model so clearly patterned after their own experiences in the American land-grant university system, they were unable to address farming populations and practices based on traditional techniques and entrenched in a social context that seemed mysterious and alien. With the exception of the educational program, the other OSS programs lacked strength for the same reason: they all required substantial negotiation with Mexican farmers who did not resemble the more traditional land-grant client. Conversely, the educational program "worked" because the Mexican participants shared a set of values and expectations similar to their American counterparts.

At issue here is the question of what is meant by "success" in foreign agricultural development. The RF, as well as independent analysts, have called the program "successful" in that it increased productivity of farm crops and created a new corps of scientific personnel capable of sustaining agricultural development. Others have criticized the program for not addressing problems of subsistence farmers, by far the largest segment of Mexico's farming population. But what both champions and critics have missed is that the land-grant system operated by means of practical information "trickling down" from research scientists to commercial farmers who were generally affluent and enthusiastic enough to adopt recommended practices. The system was kept in motion when the research was legitimated by farmer acceptance, and agricultural practice was legitimated by scientific research. The farming population in Mexico, however, was quite different from that in the U.S.A., and was unable, as a population, to assimilate the agricultural development strategies proffered by the OSS. In addition, the OSS did not aggressively generate an effective information-transmission system: the research was designed for that fraction of the Mexican farm population that most resembled commercial farmers in the U.S.A., and mechanisms for broadening the research and extension base to accommodate more farmers were not established. It is important to recognize that the OSS did designate an extension branch of their program, indicating not only their reliance on the land-grant model but also their recognition of the importance of extension as both a legitimating device and as a mechanism crucial to practical success.

Vernon Ruttan and Yujiro Hayami have argued that establishing "agricultural experiment station systems" in less developed countries is the critical step in successfully transferring "new productive capacity" from a developed to a less developed nation.[75] As I have tried to suggest, however, this model as practiced in the U.S.A. is a less than perfect fit for countries such as Mexico, whose economy, social structure, and particularly farming population, differ radically from the American standard. To be effective the model must be drastically altered to account for these differences. In addition, such alterations must be applied to the model as a whole, not merely to one or two components. The success of the land-grant model in the U.S.A. lies in its integrated, holistic approach to agricultural issues, and one might argue that any model developed for exporting agriculture will succeed by virtue of its multiple integrated functions that take into account the widely diverse features of other cultures, and that focuses not on the similarities between this culture and that, but on the differences.

Notes

1. Vernon Ruttan, "The Green Revolution: Seven Generalizations," *International Development Review* 19 (1977): 16–23; Nicholas Wade, "Green Revolution (I): A Just Technology, Often Unjust in Use," *Science* 186 (December 20, 1974): 1093–96; and Vernon Ruttan, "Green Revolution (II): Problems of Adopting a Western Technology," *Science* 186 (December 27, 1974): 1186–92.

2. Ruttan, "The Green Revolution," 20.

3. The other programs, sponsored by the Institute of Inter-American Affairs, began operating in Peru in 1943. See Arthur T. Mosher, *Technical Cooperation in Latin American Agriculture* (Chicago: University of Chicago Press, 1957), 100–126; E. C. Stakman, Paul Manglesdorf, and Richard Bradfield, *Campaigns against Hunger* (Cambridge, Mass.: Belknap Press of Harvard University Press, 1967). The best general account of agricultural development in Mexico is Cynthia Hewitt de Alcantara, *Modernizing Mexican Agriculture: Socioeconomic Implications of Technological Change, 1940–1970* (Geneva: United Nations Research Institute for Social Development, 1976).

4. Vernon Ruttan and Yujiro Hayami, "Technology Transfer and Agricultural Development," *Technology and Culture* 14 (1973): 141.

5. For a discussion of the evolving role of the RF, see Robert E. Kohler, "The Management of Science: The Experience of Warren Weaver and the Rockefeller Foundation Programme in Molecular Biology," *Minerva* 14 (1976): 279–306. The standard history of the RF is Raymond B. Fosdick, *The Story of the Rockefeller Foundation* (New York: Harpers, 1952); the most critical account of RF policy and ideology is E. Richard Brown, *Rockefeller Medicine Men: Medicine and Capitalism in America* (Berkeley: University of California Press, 1979).

6. Joseph C. Bailey, *Seaman A. Knapp: Schoolmaster of American Agriculture* (New York: Columbia University Press, 1945); Raymond B. Fosdick, *Adventure in Giving: The Story of the General Education Board* (New York: Harper & Row, 1962), 39–62; General Education Board, *The General Education Board: An Account of Its Activities 1902–1914* (New York: General Education Board, 1915). The Rockefeller Foundation Archives also have a brief description of the General Education Board program, by B. D. [unidentified], "Farm Demonstration–GEB Pro-

gram in North and South, 1906–1919," January 1, 1947. This is based primarily on the RF annual reports, since most of the correspondence regarding it has been lost. Seaman Knapp's "Ten Agricultural Commandments" were:

1. Removal of all surplus water on and in the soil.
2. Deep fall plowing, and in the South a winter cover crop.
3. The best seed, including variety and quality.
4. Proper spacing of plants.
5. Intense cultivation and systematic rotation of crops.
6. The judicious use of barnyard manure, legumes, and commercial fertilizer.
7. The home production of the food required for the family and for the stock.
8. The use of more horsepower and better machinery.
9. The raising of more and better stock, including cultivation of grasses and forage plants.
10. Keeping an accurate account of the cost of farm operations.

7. See also John Ettling, *Germ of Laziness: Rockefeller Philanthropy and Public Health in the New South* (Cambridge: Harvard University Press, 1981).

8. Fosdick, *Adventure*, 39–62.

9. It should be noted here that between 1904 and 1914, land-grant administrators and USDA officials carried on a heated dispute about the proper role of both federal and state institutions in extension efforts. State agricultural college deans felt that the southern work, extended to northern states in 1909, was an affront to states' rights and autonomy. This issue is more fully discussed in Deborah Fitzgerald, *The Business of Breeding: Public and Private Development of Hybrid Corn in Illinois, 1895–1940* (Ph.D. dissertation, University of Pennsylvania, 1985). Jim Hightower has argued that the land-grant ideal has never been realized, as evidenced in the relationship between American agribusiness and the colleges; see his *Hard Tomatoes, Hard Times* (New York: Schenkman, 1972).

10. See A. C. True, *History of Agricultural Experimentation and Research in the United States, 1607–1925* (Washington, D.C.: USDA Miscellaneous Publication 251, 1937).

11. Scholars have not yet addressed the complex interrelationships between education, research, and extension in the land-grant system in a systematic manner, although some have pointed the way; see, for example, Roy V. Scott, *The Reluctant Farmer: The Rise of Agricultural Extension to 1914* (Urbana: University of Illinois Press, 1970); Charles Rosenberg, "Rationalization and Reality in the Shaping of American Agricultural Research, 1875–1914," *Social Studies of Science* 7 (1977): 401–22; Charles Rosenberg, "Science and Social Values in 19th Century America: A Case Study in the Growth of Scientific Institutions," in his *No Other Gods* (Baltimore: Johns Hopkins University Press, 1979), 135–52; Charles Rosenberg, "Science Pure and Science Applied: Two Studies in the Social Origins of Scientific Research," ibid., 185–95; Charles Rosenberg, "The Adams Act: Politics and the Cause of Scientific Research," Ibid., 173–84; Margaret Rossiter, *The Emergence of Agricultural Science: Justus Liebig and the Americans, 1840–1880* (New Haven, Conn.: Yale University Press, 1975); Margaret Rossiter, "The Organization of the Agricultural Sciences," in Alexandra Oleson and John Voss, eds., *The Organization of Knowledge in Modern America 1860–1920* (Baltimore: Johns Hopkins University Press, 1979), 211–48.

12. Eugene Davenport, "Shall We Ask for Further Legislation in the Interest of Agriculture? If So, What?" *Proceedings of the American Association of Agricultural Colleges and Experiment Stations* (1910): 79.

13. For early solicitations for aid to Mexico, see Winton to Gates, December 10, 1913; Salmans to RF, March 15, 1915; Ramón to RF, May 28, 1915; McConnell to RF, December 15, 1915; O'Brien to RF, December 27, 1915; Winton to RF, April 19, 1916, RFA, R.G. 1.1, Series 323, Box 2, Folder 6, RAC.

14. Staff Conference, February 18, 1941, RFA, R.G. 1.2, Series 323, Box 865, Folder 1936–1941, RAC.

15. "Richard Bradfield," *American Men and Women of Science* (New York and London: R. R. Bowker for Jacques Catell Press, 1979), 1: 519; "Paul Cristoph Manglesdorf," ibid., 5: 167; "Elvin Charles Stakman," ibid., 6: 4261.

16. E. C. Stakman, Oral History, 106. All oral histories referred to were produced by William C. Cobb for the RF, and are on file at the RFA. I gratefully acknowledge the Secretary of the RF, who allowed me to study these documents.

17. Stakman, Manglesdorf, and Bradfield, *Campaigns*; "Jacob George Harrar," *American Men*, 3: 501; "Edwin John Wellhausen," ibid., 7: 517; "Norman Ernst Borlaugh," ibid., 1: 599; "John J. McKelvey," ibid., 5: 3329; "Lewis M. Roberts," ibid., 5: 5277; "William S. Colwell," ibid., 1: 1125.

18. "Agricultural Conditions and Problems in Mexico—Report of the Survey Commission to the Rockefeller Foundation, 1941," RFA, R.G. 1.1, Series 323, Box 5, Folder 33, RAC.

19. Fosdick to Gomez, October 1, 1942; Gomez to Fosdick, October 19, 1942; Gomez to Miller, February 19, 1943, RFA, R.G. 1.2, Series 323, Box 865, RAC. George Harrar, Oral History (see citation 16), 35–36, 39, 45, 182, RAC.

20. Harrar, Oral History, 55–56, RAC.

21. Stakman, Oral History, 50–51, 210–11, RAC.

22. Harrar, Oral History, 55–56, RAC.

23. Stakman, *Campaigns*, 57–58; Alcantara, *Modernizing*, 19–20.

24. Stakman, *Campaigns*; and Alcantara, *Modernizing*, 37–40.

25. Annual Meeting of the Advisory Committee, Mexican Agricultural Program, October 17, 1946, RFA, R.G. 1.1, Series 323, Box 9, Folder 56, RAC; Harrar, Oral History, 90–98, RAC.

26. Annual Meeting of the Advisory Committee, Mexican Agricultural Program, October 13, 1949, RFA, R.G. 1.1, Series 323, Box 9, Folder 56, RAC; Paul Manglesdorf, Oral History, 108–109, RAC.

27. Stakman, Manglesdorf, and Bradfield, *Campaigns*, 51.

28. Dana G. Dalrymple, *New Cereal Varieties: Wheat and Corn in Mexico* (Washington, D.C.: International Development, Foreign Agricultural Service and USDA, 1969), 19.

29. Stakman, Manglesdorf, and Bradfield, *Campaigns*, 71; Dalrymple, *New Cereal*, 6; Delbert Myren, "The Rockefeller Program in Corn and Wheat in Mexico," in Clifton Wharton, ed., *Subsistence Agriculture and Economic Development* (Chicago: Aldine, 1969), 438–52.

30. Myren, "The Rockefeller," 441. See also Dana G. Dalrymple, *The Green Revolution: Past and Prospects* (Washington, D.C.: USDA Economic Research Service and USAID Bureau in Program and Policy Coordination, 1974).

31. A. R. Mann, "Observations in Mexico, Trip Report, 9–25 April 1943," RFA, R.G. 1.2, Series 323, Box 1, Folder 4, RAC; Harrar, Oral History, 34, RAC; Alcantara, *Modernization*, 26–37.

32. Stakman, Manglesdorf, and Bradfield, *Campaigns*, 72–93.

33. Stakman to Hanson, August 2, 1944, RFA, R.G. 1.1, Series 323, Box 7, RAC.

34. Annual Meeting of the Advisory Committee, Mexican Agricultural Program, November 4, 1948, RFA, R.G. 1.1, Series 323, Box 9, Folder 56, RAC.

35. Dalrymple, *New Cereal*, 6.

36. Myren, "The Rockefeller," 444.

37. Donald Freebairn, "The Dichotomy of Prosperity and Poverty in Mexican Agriculture," *Land Economics* 45 (1969): 31–42; Dalrymple, *New Cereal*, 17, 26.

38. Dalrymple, *New Cereal*, 22.

39. Bryce Ryan and Neal Gross, *Acceptance and Diffusion of Hybrid Corn Seed in Two Iowa Communities* (Ames: Iowa Agricultural Experiment Station, Research Bulletin 372, 1950).

40. Annual Meeting of the Advisory Committee, Mexican Agricultural Program, November 4–5, 1948; October 13, 1949; October 26, 1950, RFA, R.G. 1.1, Series 323, Box 9, Folder 56, RAC.

41. This topic was poorly represented in the archival sources; the bulk of this discussion is based on Stakman, *Campaigns*, 146–49.

42. Eduardo L. Venezian and William K. Gamble, *The Agricultural Development of Mexico: Its Structure and Growth Since 1950* (New York: Praeger, 1969), 102–103.

43. Stakman, *Campaigns*, 197–215.

44. Dalrymple, *New Cereal*, 22; Mosher, *Technical Cooperation*, 119.

45. Harrar, Oral History, 25–27, 183–84, RAC.

46. Annual Meeting of the Advisory Committee, Mexican Agricultural Program, October 15, 1945; October 17, 1946; October 30–31, 1947; October 13, 1949, RFA, R.G. 1.1, Series 323, Box 9, Folder 56, RAC.

47. Annual Meeting of the Advisory Commitee, Mexican Agricultural Program, November 4, 1948, RFA, R.G. 1.1, Series 323, Box 9, Folder 56, RAC.

48. Annual Meeting of the Advisory Committee, Mexican Agricultural Program, October 15, 1945, RFA, R.G. 1.1, Series 323, Box 9, Folder 56, RAC.

49. Annual Meeting of the Advisory Committee, Mexican Agricultural Program, October 17, 1946; October 30–31, 1947; November 4–5, 1948; October 26–27, 1950, RFA, R.G. 1.1, Series 323, Box 9, Folder 56, RAC.

50. See citation 6.

51. Dalrymple, *New Cereal*, 22.

52. For example, the Illinois agricultural experiment station required a substantial demonstration of farmer interest before setting up extension work; see Eugene Davenport to the Illinois Association of Agricultural Advisors, Agricultural Series 8-21-21, Box 1, File 1914, University of Illinois Archives.

53. See Mann, "Observations," 7–8.

54. Weaver to Harrar, March 19, 1948, RFA, R.G. 1.2, Series 323, Box 5, Folder 32, RAC.

55. Ibid.

56. Weaver to John D. Rockefeller, III, October 11, 1946, RFA, R.G. 6, Series 915, Box 3, Folder 20, RAC.

57. See Agriculture, Scholarships, 1946–1955, RFA, R.G. 1.1, Series 323, Box 8, Folders 43–52. These folders contain dossiers on each student sent abroad during this period.

58. Wellhausen to Harrar, May 28, 1954, RFA, R.G. 1.1, Series 323, Box 7, Folder 41, RAC. The letter contains a listing of all RF scholars and fellows of this period, including the position assumed in Mexican agriculture.

59. Harrar, Oral History, 22–24.

60. For a description of Chapingo in 1943, see Mann, "Observations," 6–7.

61. Harrar, Oral History, 57–58, 166, 161–62.

62. Stakman, Manglesdorf, and Bradfield, *Campaigns*, 177–96.

63. Annual Meeting of the Advisory Committee, Mexican Agricultural Program, November 4–5, 1948; October 13–14, 1949, RFA, R.G. 1.1, Series 323, Box 9, Folder 56, RAC; Wellhausen to Harrar, August 13, 1953, RFA, R.G. 1.1, Series 323, Box 7, Folder 41, RAC.

64. Annual Meeting of the Advisory Committee, Mexican Agricultural Program, October 26–27, 1950, RFA, R.G. 1.1, Series 32, Box 9, Folder 56; Wellhausen to Harrar, August 22, 1953, RFA, R.G. 1.1, Series 323, Box 7, Folder 41, RAC; "Memorandum Supplementing Minutes of Meeting of Advisory Committee," October 26–27, 1950, RFA, R.G. 1.1, Series 323, Box 9, Folder 56, RAC.

65. Harrar, Oral History, 69.

66. Annual Meeting, October 15, 1945, 1–2, RFA, R.G. 1.1, Series 323, Box 9, Folder 56, RAC.

67. Stakman, Manglesdorf, and Bradfield, *Campaigns*, 187–89.

68. Harrar to Wellhausen, April 23, 1953; Harrar to Wellhausen, June 5, 1953; Wellhausen to Harrar, August 11, 1953, RFA, R.G. 1.1, Series 323, Box 2, Folder 6, RAC.

69. Extract from Weaver's Diary, January 14–21, 1952; Harrar to Garza, August 1952; all in Agricultural Education, Monterrey, 1952, Box 874.

70. "List of Persons Who Have Been Trained in Rockefeller Foundation, Mexico," RFA, R.G. 1.2, Series 323, Box 7, Series 41, RAC.

71. Actually, Harrar had felt prepared to begin helping Colombia as early as 1946, but the rest of the advisers did not. See Harrar, Oral History, 83–84.

72. Stakman, Manglesdorf, and Bradfield, *Campaigns*, 216–17.

73. Ibid.

74. Annual Meeting, October 1951, Project 25, RAC.

75. Ruttan and Hayami, "Technology Transfer," 132.

5

The Rockefeller Foundation's Mexican Agricultural Project

A Cross-Cultural Encounter, 1943–1949

Joseph Cotter

MANY SCHOLARS HAVE considered the establishment of the Rockefeller Foundation's Mexican Agricultural Project (MAP) in 1943, and the Green Revolution that its scientists created, as an encounter between two cultures that produced negative results for Mexico, especially for the peasant farmer.[1] In general, academics have seen the program from the perspective of the 1960s, '70s, and '80s, after farmers in many third-world countries had adopted the technologies developed or promoted by the RF's scientists, and the socioeconomic consequences of the Green Revolution were evident. Previous historical studies have relied on archival sources available in the United States. Examining these matters through research in Mexican archives brings new insight to the years when the MAP was a young and novel program.

In several areas, the RF's personnel demonstrated keen insight into the needs and desires of the relevant Mexican actors. In others, they could not bridge the gap between cultures. In general, during the 1940s the Mexican government, agricultural science community, media, and farmers supported the MAP and its research.[2] This chapter discusses the origins of the MAP, U.S. and Mexican views on science and agricultural modernization, the RF's understanding of Mexican cultural norms, and the reception of the MAP in Mexico.

The 1940s were critical years for Mexican agriculture. The Mexican state shifted from the proagrarian policies of President Lázaro Cárdenas (1935–40) to an emphasis on urban industrialization under the more conservative President Manuel Avila Camacho (1941–46). Finding technical solutions to the problems of Mexican agriculture became more important than during previous administrations, when such programs were often closely linked to reformist measures to improve the lot of the rural poor.

Previous works have seen the origins of the MAP as a reflection of the new orientations of the 1940s.[3] However, historical trends and events in Mexico before 1940 played a significant role in setting the stage for the MAP's arrival. From the Mexican perspective, the establishment of the MAP was indeed timely. Twenty years of ineffective research and extension efforts, and shortages of basic foodstuffs during the late 1930s and early 1940s, made the success of programs to improve the technical aspects of farming a pressing issue and stimulated the government's interest in the MAP and other U.S. technical-assistance programs for agriculture. During the 1940s, providing farmers with new and improved inputs derived from scientific research—something previous administrations had tried but failed to accomplish—helped enhance the image of competence of both the state and the agricultural science community.

Further, rightist attacks that blamed the land-reform sector for the low productivity of Mexican agriculture forced the government to do something politically to "save" the agrarian reform. Criticism of land-reform always surfaced in times of agricultural crisis; deflecting it by applying scientific research to staple crops was a logical response. The government followed this approach in the early 1930s, employing Mexico's own agricultural scientists. During the late 1930s and early 1940s, critics of land reform became vocal again, and the government hired scientists from the United States to conduct research. Even if the land-reform sector often did not benefit from the MAP's programs, the increased productivity and image of scientific success that the Green Revolution produced helped to defuse attacks on the agrarian cause. By 1941 the government was concerned less with land reform and social justice for the peasantry, and more with the immediate shortages of farm products it faced; hence, it did not worry about who would benefit from the MAP's research when it decided to accept the Rockefeller Foundation's aid.

During the 1940s, the Mexican agricultural science community was also worried more about the immediate problems it faced than about its commitment to the cause of the peasantry. Its previous failure to increase the productivity of agriculture, its close association with the agrarian reform, and its low credibility with Mexican farmers and the general public made immediate action to diffuse attacks and improve its prestige essential. By 1941, the agricultural research community, as well as other segments of society, felt the time for revolutionary rhetoric and agrarian reform had passed. The moment had come to transform the "agrarian revolution" into an "agricultural revolution."

The members of the Mexican agricultural science community realized that they could not do this on their own and gladly accepted the aid of the RF and its scientists. The foundation understood their desire to change their public image, and developed policy for the MAP accordingly. The MAP's early years were a cross-cultural encounter that benefited both the Mexican government and the Mexican agricultural science community.

The Origins of the Green Revolution: A Mexican Perspective

The formation of the Mexican agricultural research community began in 1864, when the government established the Escuela Nacional de Agricultura (ENA). Until the end of the Porfirio Díaz dictatorship (1911), the ENA's curriculum was designed to prepare administrators and advisors for Mexico's haciendas.[4] During the 1920s and 1930s, the school stressed the theoretical aspects of agriculture and provided little training in the methods of experimental science.[5] At the same time, in reaction to Mexico's reliance on foreign research and agricultural technologies, and in keeping with the revolutionary slogan *¡México para los Mexicanos!*, some members of the agricultural science community adopted an ideology of "scientific nationalism": the belief that, to be successful, programs to modernize agriculture had to be based on research done in Mexico by scientists familiar with the social goals of the revolution.[6] The lack of an experimental tradition in Mexican agricultural science mitigated against this autarkic approach to research, and from 1917 to 1943 had a profound effect on agricultural science policy.[7]

When the outbreak of the Mexican Revolution forced the ENA to close, many of its students joined the agrarian cause. The vast majority of Mexico's agronomists supported land reform in the 1920s and 1930s; some were among its most vocal advocates.[8] They also participated in its implementation. In 1935, when Cárdenas accelerated the pace of land distribution, the number of agronomists employed at this task grew vastly, to outnumber those working in research and extension programs.[9] Further, many agronomists served in political positions.[10] By 1940, the public closely associated the profession of agronomy with the politics of the government, especially the agrarian reform.

During the early 1920s, the agronomists spoke of the need to modernize Mexican agriculture and recommended that the state employ more members of the profession to accomplish this end.[11] The agronomists' vision of modernizing agriculture included not only improving the technical aspects of farming but promoting better rural hygiene, organizing farmers to bring them economic empowerment, and raising the political consciousness of the peasantry. The agronomists wanted to tackle all of Mexico's many rural problems simultaneously. Until the 1940s this notion, labeled *proyectismo* by Eyler N. Simpson, formed the foundation of the scientific community's approach to modernizing agriculture.[12]

During the Venustiano Carranza administration (1915–20), the Mexican government began to support agricultural research and extension programs. In keeping with the lack of an experimental tradition, however, extension and agricultural education took precedence over research.[13] To improve corn cultivation, the Secretaría de Agricultura y Fomento (Ministry of Agriculture and Development, SAF) developed the *campaña en pro del maíz* (campaign in favor of corn), using lectures, demonstrations, radio broadcasts, and articles in the national press to convince farmers to select their seed from standing plants with certain morphological characteristics, disinfect it with commercial products, store it properly, and conduct germination tests to ensure its viability. Although the Mexican agricultural science community was familiar with the accomplishments of corn breeders in the United States, its members lacked experience with and general interest in experimental science; they expected the farmer, rather than the agronomist, to do the plant breeding.[14]

In 1930 the onset of the Great Depression, and poor corn harvests that forced the government to resort to imports, sparked much debate in Mexico regarding the agrarian reform and the modernization of agriculture. Some prominent intellectuals and political leaders claimed that land reform was the cause of Mexico's low agricultural productivity, and accused the agronomists of accomplishing nothing to improve it.[15] The agronomists in turn charged that the government did not give adequate support to research and extension efforts. They pointed to internal problems in the SAF—poor planning and leadership, and lack of continuity in programs—as the reasons for failures.[16] Reiterating the theory of "scientific nationalism," the agricultural science community recommended that more resources be devoted to research, and continued to advocate the *proyectismo* approach.[17] According to agronomist Manuel Mesa A., limiting the profession's actions to "the simple execution of technical works," as some advocated, would not lead to the improvement of Mexican agriculture, but rather would "condemn all that has happened in agricultural policy up to the present."[18]

In 1933 and 1934, to increase its support for research, the government reformed its agricultural science policy, and established the Instituto Biotécnico, a central research institute, as well as regional experimentation stations.[19] The SAF used both research and extension programs to attempt to increase the productivity of many different crops, and devoted less attention to corn than during the 1920s. The research projects accomplished little, however, in terms of applications that could improve the productivity of farming, and few extension programs were based on a solid foundation of research. Most of the research projects were actually empirical tests of foreign improved crop varieties, such as "Mentana" wheat which was created in Italy.[20] The government continually increased the number of experimentation stations, despite the fact that inadequate resources limited their operational effectiveness.[21] The image of science serving the peasantry was more important than scientific achievement.[22]

In 1933 Mexico became self-sufficient in corn, giving rise to optimism with regard to programs to improve agriculture. In 1937 Cárdenas, a spokesman for the SAF, and members of the agricultural science community publicly praised the success of previous campaigns and claimed that the ministry had made progress toward solving the problems of Mexican corn growers.[23] However, that same year, shortages of maize occurred in various locations, and by 1938 imports were required to meet demand. This reversal destroyed the euphoria of the government and research community and led to some popular discontent.[24] The SAF responded by resurrecting the *campaña en pro del maíz*, this time making its practices mandatory in regions under the influence of its extension agents. The campaign reached roughly 6 percent of Mexico's cornfields and did not alleviate the need for further imports.[25]

During the late 1930s, the corn crisis, a plague that threatened to destroy Mexico's banana industry, and problems with other crops led to more criticism of the agrarian reform and agronomists.[26] Further, many farmers continued to demand that the government provide them with technical support.[27] Some agronomists believed that the situation was hopeless. For example, according to Miguel García Cruz,

> Agricultural and industrial experimentation is very deficient, they do not give us the means to improve it, and when they have the use of some resources, the strong political pressure makes them fail. Our laboratories and experimentation stations do not have the resources, nor the technical personnel with the stability and responsibilities needed for this worthy program.[28]

The profession's prestige descended to a new low, and people both inside and outside the research community began to argue that Mexico needed foreign scientific assistance to solve its agricultural problems.[29] For the duration of Cárdenas's presidency, the SAF did not change its approach to research and extension programs and continued to expand the number of operating experimentation stations—regardless of supply shortages and the lack of personnel with training in experimental science—which ensured that they could not conduct useful research.[30]

In 1941, to avoid the mistakes of the past, agronomist Marte R. Gómez, the new Minister of Agriculture, implemented reforms in the ministry. Many members of the agricultural science community recognized that an effective research program was essential to modernizing Mexican agriculture.[31] To put the SAF's programs on a more scientific foundation, Gómez disbanded the extension service and assigned its personnel to the experimentation stations, while at the same time reducing the number of stations in operation, thus concentrating efforts to facilitate the success of research projects.[32] At the suggestion of U.S. Vice-President Henry A. Wallace (reinforced by the scientists of the RF's site-survey team, and the results of a similar study by personnel from the United States Department of Agriculture), the SAF began a plant-breeding program that concentrated on corn and beans.[33]

Although Gómez's reforms gave experimental science more support, they still struggled with the legacies of the past. Research projects were more successful than those of previous years but suffered from shortages of trained personnel and lack of continuity in programs. Inadequate funding handicapped various SAF programs and had an especially adverse impact on pest-control campaigns. On occasion, politics and personal considerations still influenced the agency's projects.[34] By 1946, despite Gomez's desire to bring the number of operating experimentation stations in line with budgetary realities, populism, farmers' demands for scientific support, and the need for increased agricultural production due to the war and Avila Camacho's industrialization program, had forced the SAF to establish seven new stations.[35] However, from 1944 to 1946, insufficient financial support caused the stations to deteriorate to the point where J. George Harrar, the director of the MAP, felt that this issue should be among the first the RF raised with President-elect Miguel Alemán (1947–52).[36]

Gómez's reforms did not silence the agronomists' critics. For example, a poultry farmer belittled their scientific abilities, and wrote, "their studies were not related to the exploitation of the land, but rather of the national treasury."[37] Various individuals lambasted the agronomists for their excessive participation in politics, as well as their lack of interest in increasing the productivity of farmers.[38] Rightist groups such as the *Sinarquistas*, which had a large following among certain segments of the peasantry, as well as private farmers with large holdings, had a special contempt for the profession because of its advocacy of land reform.[39] Few Mexican farmers, even land-reform beneficiaries, had much faith in the SAF or in the agricultural research community's abilities as scientists or technical advisors.[40]

Production of corn was once more insufficient to satisfy national demand in 1943, and Gómez had to tell Avila Camacho that the SAF was not yet capable of doing anything to address the problem.[41] In lieu of providing farmers with new, high-yielding corn varieties, or undertaking campaigns to raise yields by means of increasing the use of fertilizers or improving cultivation practices, the government had to force all farmers to plant at least 10 percent of their land in corn, an unpopular measure among those that cultivated more-lucrative crops such as cotton.[42] In 1945 Gómez was accused of using his position as Minister of Agriculture for personal gain.[43] Gómez, who was especially concerned about his profession's prestige, desperately wanted a scientific success that would improve the agronomist's public image.[44]

In February 1943, after two years of anxious waiting while the foundation searched for appropriate personnel, Gómez and President Avila Camacho signed the agreement that created the MAP.[45] The RF appointed Harrar, a plant pathologist, as the project's field director in Mexico.[46] The agreement that established the MAP was one of several Mexican attempts to seek U.S. technical aid for agriculture during the late 1930s and early 1940s.[47] The failure of previous research and extension programs, and the agronomists' desire to improve their public image, set the stage for Mexico's quest for foreign scientific aid in agriculture during the 1940s, and hence, the arrival of the MAP.

The MAP and Mexican Views of Agricultural Science

From 1941, when the RF's site-survey team toured Mexico, until the end of the decade, foundation personnel believed that Mexican agricultural science was backward, and stressed the need to improve the research community's knowledge of experimental methods.[48] These notions meshed well with the goals of that community. The agronomists' desire to make their profession a scientific one did not arise entirely as a result of the events of the 1940s, but rather first emerged during the 1920s and early 1930s in struggles to make the curriculum at the ENA less "encyclopedic" and to introduce specializations in the profession of agronomy.[49] During the mid-1930s, several agronomists lamented the profession's deficient knowledge of experimental science, and by the end of the decade the movement to transform the profession into a scientific one was gaining momentum.[50]

By the early 1940s, the majority of the agronomists agreed with the RF's assessment of their abilities as research scientists. Marco Antonio Durán, an agronomist who advocated making his profession more scientific, told the April 1941 meeting of the Sociedad Agronómica Mexicana (Mexican Agronomic Society, SAM):

> We have distracted our attention in aspects of land distribution and economic organization, the time has come for at least a segment of the agronomists to adopt a completely scientific attitude, which would also bring prestige to agronomy.[51]

Senior agronomists sought to enhance the profession's familiarity with the methods of experimental science, and students at the ENA wanted to learn them.[52]

In addition to stressing the agency's new focus on research, the SAF's annual reports denounced the previous linkage of the technical modernization of farming with social, economic, and political issues, and proclaimed that Mexican agricultural science had been "depoliticized." The SAF's *Memoria* for 1945 and 1946 stated:

> In a matter that was traditionally expressed as political, and relying on the largest budget in Mexico's history, we have positively abstained from taking part in political debates and from putting financial resources or the influence of the Offices of the agency in first, second, or third person, to the service of whichever political interest.
>
> Good or bad, we do agricultural work, exclusively agricultural. It is oriented toward the technical solution, apolitical, at least in the way that politics connotes personalism and meanness, of the rural problems of Mexico, without contacts with factions, without compromising in the matter of principles, without trying to deviate toward other goals . . . [53]

Bruce Jennings argues that the MAP played a critical role in the depoliticization of Mexican agricultural science, similar to what the RF accomplished during the early twentieth century with medical science in the United States.[54] Although the project contributed to the process, however, the roots of depoliticization predate the MAP's arrival in Mexico.[55] During the late 1930s, various Mexican agronomists (interestingly, especially those on the political left) argued that agricultural research and extension programs should be depoliticized and separated from the struggle for economic justice.[56]

However, many members of the agricultural science community were still interested in economic, social, and political issues during the 1940s, whereas some believed that the technical improvement of agriculture should be separated from these concerns.[57] Gabriel Leyva Velázquez, the head of the Confederación Nacional Campesina (National Peasant League, CNC), exhorted the agronomists not only to think of themselves as technical experts but also to continue to support the linking of social, economic, and political issues with the modernization of agriculture.[58] In 1945 a conflict among the agronomists led to a schism in the SAM. Debate over whether the agronomists should support the candidacy of Miguel Alemán, who was unsympathetic to the agrarian cause, or form closer relations with the CNC and advocate more land reform, contributed to the dispute. Neither faction, however, attacked the MAP.[59]

By the mid-1940s, most members of the agricultural science community agreed that improving the productivity of farming was a more pressing need than solving the other problems of rural Mexico. That is, the other problems could not be solved without first modernizing Mexican agriculture.[60] Influential members of the profession lamented the ineffective efforts of the 1920s and

1930s and stressed the urgency of improving agricultural-modernization pro-grams.[61] In response to Avila Camacho's call to the agronomists to make agri-culture the "foundation of industrial greatness," they acknowledged the need to concentrate on improving the technical aspects of farming by means of scientific research.[62] An August 1941 position paper by the SAM faction that in 1945 tried to seize control of the society in the name of the peasantry stated,

> The agronomists associated with the society have as a principal goal putting their knowledge to the service of solving the agricultural problems of the country, divested completely of interests of a political nature because they consider that in the present moment, the most pernicious thing for the direction of national agriculture as likewise for the life of our profession and for the best activity of each one of the agronomists, is the link that could be established in our function as technical experts and professionals with political ups and downs (*arribismos*).[63]

Many land-reform beneficiaries and private farmers supported the president's call to improve the productivity of agriculture.[64]

It would seem obvious that rightist groups and private farmers would lead the movement to depoliticize agricultural science. However, a representative of the Liga de Comunidades Agrarias (League of Agrarian Communities) of the state of Veracruz (who was also a member of the CNC) told Avila Camacho, "It is undeniable that the technical knowledge of said professionals [the agron-omists] will be a factor of indisputable worth" in the improvement of Mexican agriculture, "when they divest themselves completely of all matters of a fac-tional order, and dedicate themselves entirely to their technical labors."[65] José Ch. Ramírez, Secretary of the Camara de Diputados (Chamber of Deputies), demanded that the SAF focus solely on the scientific aspects of agricultural modernization.[66]

In reaction to the profession's low prestige and in keeping with the broader change in emphasis from political to technical solutions to national problems, by the early 1940s a large number of agronomists, particularly the younger members of the profession, wanted the public to think of them as scientists rather than as bureaucrats or agrarian agitators.[67] Although some conflict arose among the agronomists over which sector of rural Mexico would be targeted by the government's programs, no disagreement existed concerning the need for the state and the agricultural science community to improve farm produc-tivity through research. Hence, during the 1940s, proagrarian agronomists did not attack the MAP.[68] The depoliticization of Mexican agricultural science did not take place solely as the result of the imposition of an apolitical approach by the MAP.

The MAP and Models of Farming

The original research plans for the MAP, which were determined after negotiations with the SAF's directors, targeted four areas: the control of wheat
rust, the improvement of the SAF's corn-breeding program, studies related to
the management of soils, and research on various livestock diseases.[69] Mutual
agreements between the RF and SAF later added the varietal improvement of
beans, the introduction of soybean cultivation, and studies of forage crops. In
1943 research began on the development of hybrid corn and rust-resistant
wheat varieties, and soil-management studies were initiated in 1945, but various problems delayed the research on animal diseases. The MAP also studied
chemical methods of controlling agricultural pests, and appropriate chemical
fertilizer formulas for various crops. On occasion MAP personnel worked with
representatives of U.S. agribusiness firms, testing and promoting the firms'
products, such as insecticides and farm machinery.[70] The Rockefeller Foundation's program included sending promising Mexican students in the agricultural sciences to the United States for graduate study.

The incoming Alemán administration implemented further reforms in the
SAF in 1947, creating the Instituto de Investigaciones Agrícolas (IIA), a research organization directed and operated by Mexican scientists, and the
Comisión Nacional del Maíz (National Corn Commission, CNM), an autono-
mous agency that was responsible for distributing the new corn varieties to
farmers. Problems similar to those of previous SAF research efforts, such as
underfunding, overextended programs, and rapid job turnover, limited the effectiveness of the IIA.[71]

By 1948 MAP scientists had several hybrid corn and wheat varieties ready
for distribution to Mexican farmers. The plants required the timely application
of chemical fertilizers and pesticides to reach their highest potential yields,[72]
but from 1950 to 1965, the productivity of agriculture increased dramatically,
forming part of the Mexican "economic miracle."[73] During the late 1940s, the
CNM controlled the distribution of the new corn seeds, despite the reserva-
tions of RF personnel, who believed that the organization was overly political
in nature.[74] The distribution of the MAP's hybrid wheats remained under the
control of the SAF and proceeded more efficiently.[75] Although fewer corn
growers than wheat growers adopted the new varieties, by retaining control of
the distribution of the new plants, the Mexican government and agricultural
research community benefited from the political capital generated by the results of MAP research.[76]

In a reflection of the 1945 schism in the Sociedad Agronómica Mexicana
over the profession's relationship with the peasantry, the IIA promoted "open-

pollinated" corn varieties, which were believed to be more appropriate for the conditions of peasant agriculture because farmers could use some of their harvest for seed, as they traditionally did with native corn varieties, rather than having to purchase new seed annually, as must be done with the RF's hybrids for them to reach their full potential.[77] Relations between the IIA and the MAP were strained at times, but during the 1940s this did not lead to a widespread reaction against the MAP among Mexican agricultural scientists.[78] The majority of the most prominent agronomists were not IIA researchers and hence did not think of the MAP scientists as professional rivals.

Deborah Fitzgerald argues that the MAP's research represented the imposition of a U.S. model of agriculture on the radically different conditions of rural Mexico. Hence, the farmers who were most like U.S. farmers, especially the wheat growers of northern Mexico, were able to benefit from MAP technologies, while the small-scale, peasant corn cultivators could not.[79] The MAP's research and the technologies it created certainly paralleled the methods of scientific farming developed by U.S. land-grant colleges and were not oriented toward meeting the needs of the peasant farmer. Further, the RF received several warnings from academics such as Carl O. Sauer about the dangers of transplanting the methods of U.S. agriculture to Mexico, and several of the MAP's scientists recognized the value of the technologies and methods developed over the centuries by Mexico's peasants, but chose not to study them in detail or to incorporate them into their research program.[80]

Although the model of agriculture offered by the MAP ignored the methods (and often realities) of peasant farming in Mexico, the majority of the Mexican agricultural science community also did not believe that those agricultural practices were effective or worthy of study.[81] For example, some Mexican agricultural scientists (as well as the SAF) attacked the traditional peasant practice of intercropping corn and beans.[82] After recommending that Mexicans import Rhode Island Red roosters to improve the poultry industry, Rufino Monroy, during the 1920s the SAF's specialist in this field, wrote, "The principal enemy of poultry keeping in Mexico is the Indian with his native rooster."[83] However, the scientists of the IIA demonstrated more concern over bringing the benefits of research to the peasant farmer than did the MAP's corn breeders.[84] During the 1940s, many members of the Mexican agricultural science community advocated the use of methods to modernize agriculture similar to those of the MAP's researchers. Although some recognized that MAP technologies would not be appropriate for the nonirrigated lands of most Mexican farms, at this point these individuals did not criticize the project.[85]

Further, U.S. agricultural scientists presented the Mexican government with alternative approaches to the improvement of farming that were rejected. After a 1944 study of Mexican agriculture under the auspices of the Pan American Union, ecologist William Vogt prepared a report for the government that proposed that farming in Mexico be modernized by improving soil-conservation practices and promoting nonchemical methods of pest control.[86] During the early 1940s, some in Mexico advocated government programs to improve soil-management practices, and Avila Camacho established a soil-conservation service as a branch of the National Irrigation Commission.[87] However, during the Alemán administration, the SAF gave only limited support to a MAP program to promote the use of green-manure crops to improve soil fertility.[88] In 1950, when J. I. Rodale, an advocate of organic farming methods, came to Mexico to promote alternatives to the use of chemical fertilizer, the SAF was not interested in his ideas.[89]

Even before the 1940s, some Mexicans wanted to emulate the "scientific" agriculture of the United States. During the 1920s and 1930s, many Mexicans admired U.S. farming practices and agricultural research and extension programs.[90] Experiments with chemical fertilizers were among the first research programs conducted by the SAF.[91] By the late 1930s, although some members of the Mexican agricultural science community advocated organic methods of enhancing soil fertility, most were convinced that increasing the use of chemical fertilizer was essential.[92] In 1938 and 1939, Cárdenas considered financing the construction of a nitrate fertilizer factory, but he decided against it, probably because of its cost.[93] From 1926 to 1937, the SAF gave limited support to agronomist Pandurang Khankhoje's plant-breeding experiments with corn.[94]

In 1941–42 the SAF began a series of experiments to find the best formulas of chemical fertilizers for sugarcane and established a new laboratory to tighten controls on the contents of pesticides and fertilizers.[95] The Mexican government committed to the improvement of agriculture by means of promoting the use of chemical inputs when it established the parastatal firm Guanos y Fertilizantes de México in 1943.[96] During the 1940s, most members of the Mexican agricultural research community—like the MAP scientists—wanted Mexican farmers to use more chemical fertilizers and pesticides.[97] The Alemán administration was especially committed to promoting chemical fertilizers. By 1949 Guanos y Fertilizantes was producing more product than it could sell, and the fertilizer requirements of the new crop varieties released by the MAP provided a market for the surplus.[98] Although the model of farming presented by the MAP influenced the outcome of the Green Revolution in Mexico, the policy of the government, as well as the attitudes of the Mexican agricultural science community, also had an important effect on the post-1943 transformation of Mexican agriculture.

Many Mexican farmers, including some in the land-reform sector, asked the government to provide them with the sort of technical support and systems of farming that the MAP offered. For example, to facilitate their availability, the CNC and several groups of land-reform beneficiaries asked the government to support a national chemical fertilizer industry.[99] The CNC rejected the *campaña en pro del maíz* approach to the improvement of corn cultivation and argued that scientists working for the government must conduct "scientific works of high research."[100] During Alemán's campaign for president, the land-reform sector and private farmers agreed that the government should give priority to introducing improved agricultural techniques.[101] Increasing production was a more pressing issue than emancipating the peasantry.

RF Publicity Policy and the Mexican Agricultural Science Community

The exigencies of World War II led to a large expansion of activity by U.S. agricultural scientists in Mexico, under the auspices of the United States Department of Agriculture and other government agencies. In response to the likely cutoff of East Asian sources due to Japanese military activity, the U.S. and Mexican governments initiated a joint program to promote and improve the cultivation of *Hevea brasiliensis* rubber trees, and numerous scientists from the United States roamed the Mexican countryside in search of wild plants that could be sources of the strategic material. Cooperative campaigns were undertaken to promote the cultivation of other crops that provided raw materials essential for the war effort, such as castor beans (the oil from which was used as an industrial lubricant). These projects received much coverage in the Mexican press.[102] Despite the negative outcomes of some of these programs, they were generally well received by the Mexican agricultural research community.[103] However, the high-profile presence of U.S. scientists in Mexico irritated some of its members.[104]

Unlike the U.S. government, the Rockefeller Foundation decided to minimize the publicity connected with its activities. Although both the MAP and the USDA's programs were established partly out of the desire to improve U.S.-Mexican relations, the RF's previous experience with technical-assistance projects in Latin America helped its personnel recognize the important role that publicity played in accomplishing the foundation's objectives, as well as in influencing the local reception of such campaigns.[105] The RF understood the occasionally nationalistic attitude of the Mexican agricultural research community—and most significantly the desire of that community to improve its public image—and structured policy regarding MAP publicity accordingly.[106] From the RF's first expression of interest in providing technical assistance for Mexican agriculture, the foundation's directors and the MAP's scientists agreed that allowing the SAF and Mexican researchers to take most of the public credit for the project's achievements best served the interests of the project.[107] Hence, the RF let the SAF issue all public statements discussing MAP activities.

The SAF used this situation to its advantage by stressing the agency's successes in agricultural research, often downplaying or neglecting to mention the contributions of the MAP scientists.[108] This did not bother the RF researchers, who only urged caution concerning the SAF's sometimes highly optimistic press releases. During the early years of the project, the RF also tried to minimize publicity about it in the U.S. press.[109] Seeking to increase the Mexican research community's prestige, the RF maintained this policy through the early years of the Alemán administration, and did not publicize the MAP's activities on its own until 1949.[110]

The MAP and Mexican Culture

Although the RF directors and the MAP scientists recognized the desires of the Mexican research community, developed their publicity policy accordingly, and provided a model of farming that many Mexicans advocated, they did not have a similar understanding of certain other aspects of Mexican culture. In particular, they had difficulty coping with the Mexican notions of *personalismo* and patron-clientelism.

The foundation's fellowship program, which sent promising Mexican students of agronomy to the United States for graduate study, was the source of the conflict. The attitude of many returning RF fellows troubled Elvin C. Stakman. He complained of "the tendency of some of the young Mexicans to expect too much in the way of emoluments and special privileges," and called these "unreasonable expectations" that "must tax the patience of even the most altruistically minded individual."[111] In 1948 he wrote, "Some of the fellows think the *Oficina de Estudios Especiales* [the MAP] should be a benevolent agency to satisfy their desire for personal and lucrative activity," and

> the expectations of some of the returned fellows are preposterous.
> Some of them seem to think that when the Foundation gave them a
> fellowship it also assumed the obligation of an indulgent foster parent.
> The 'You gave me a bathing suit, now dig me a lake to swim in'
> attitude is too prevalent. And when the lake is provided, the
> temperature of the water must be then statistically controlled.[112]

Clearly, Stakman did not understand, or could not accept, the nature of Mexican patron-clientelism, in which benefactors aid younger individuals for life in return for their support, which forms the patron's *camarilla* (network) of clients.[113]

Both Stakman and Edwin J. Wellhausen, the MAP's corn breeder, had trouble dealing with the important role personalism plays in Mexican organizations. Stakman wrote of the Mexican agronomists that they still think too much in terms of special favors granted by friends rather than in terms of sound and stable institutional organizations to which they must adjust themselves and which they must help develop.[114]

Wellhausen had problems with the attitude of Mexican researchers toward cooperative work.[115] The U.S. scientists' notions of proper organizational behavior differed from those of the Mexican agricultural science community.

The Reception of the MAP in Mexico during the 1940s

During the first years of the MAP's existence, although its personnel had some problems dealing with Mexican cultural norms, their understanding of the desires of the Mexican agricultural science community, the project's success at developing new agricultural technologies in a timely fashion, and its presentation of a model of farming that many in Mexico wanted to emulate contributed to the popularity of the program. Most members of the Mexican agricultural science community supported the MAP during the 1940s. Agronomists from the Antonio Narro Agricultural College asked the RF to provide assistance for its research program, similar to that given to the ENA.[116]

Ramón Fernández y Fernández praised the MAP, and argued that its autonomy from the SAF could be the key to its success, for this would allow it to avoid the deleterious influence of the ministry's internal politics and bring needed continuity to research efforts.[117] Marco Antonio Durán praised the project in his 1947 monograph that exhorted the Mexican government to make farming more productive.[118] In a 1950 report on Mexican agriculture, agronomist Manuel Mesa A. gave the MAP a glowing review. He wrote,

> The research and works relative to the improvement of seeds,
> principally due to the cooperation of an institution like the Rockefeller,
> which relies on valuable technical resources, and to that program of
> works that has been maintained without variation for several years,
> has brought as a result that Mexico now has not only various varieties
> of improved seeds, but also technical personnel with the capacity to
> continue researching that which before had been studied deficiently.[119]

The scientists of the MAP attended meetings of the SAM, and from 1944 to 1946 Harrar even served as one of Mexico's representatives to the Joint U.S.-Mexican Agricultural Commission.[120].

In general, during the 1940s, the Mexican agricultural science community gratefully accepted any technical assistance offered to Mexico by U.S. agencies.[121] Most of its members felt that such aid would be helpful to Mexico. For instance, the pro-agrarian wing of the SAM stated:

> The scanty demographic density of our country justifies that the
> Republic would open its doors to foreign elements kindred and able to
> assimilate to our population, divested of humiliating superiority,
> without racist complexes, nor covetous of anti-social privileges, that
> would come to our country to give creative impulse to agriculture,
> industries, science, and art, furnishing their efforts, their capital, or
> their scientific property. . . . [122]

Many in Mexico favored greater international cooperation in the sciences, and
abandoned previous notions of scientific nationalism.[123]

The agronomists appreciated the prestige they gained as a result of the
MAP's scientific successes.[124] Veteran members of the profession could devote
their attention to discussions of the relative merits of further agrarian reform,
agricultural credit problems, and other topics closer to their own expertise, and
leave experimental science in the competent hands of MAP scientists.[125] The
search for technical solutions to the problems of Mexican agriculture had become more important than the question of who would benefit from them.

During the 1940s, other groups in Mexico also supported the MAP. Most
articles in the Mexican press praised the program.[126] In 1949 the governors of
five Mexican states visited the MAP, along with many other "distinguished and
influential persons who are interested in soliciting RF collaboration."[127] As one
might expect, farmers with large holdings supported the MAP.[128] However,
many small farmers and land-reform communities also expressed interest in
the MAP's activities and sought to participate in its experiments or to obtain
the hybrid corn it released.[129] In one case, representatives of the peasant organizations of the state of Veracruz asked the government for hybrid corn seeds
again, even though the seeds had done poorly the previous year.[130] In general,
the MAP was a popular program.

Serious friction between the MAP's scientists and the Mexican agricultural
science community did not emerge until the early 1950s, after the latter had
reaped many benefits from the project's early achievements.[131] This conflict
notwithstanding, some agronomists continued to speak favorably of the MAP's
contributions to Mexican agricultural science.[132] In 1968–69, during the heady
years of the Green Revolution, the Gustavo Días Ordaz administration (1964–
70) took pride in Mexico's contribution to this demonstration of the wonders
of modern science.[133]

Conclusions

Decrying the unsuitedness of the technologies to the farming conditions
common to the third world's peasantry, many scholars have seen the technologies of the Green Revolution as an imposition of the science, agricultural prac-

tices, and culture of the developed nations onto those of the third world.[134] In contrast, the in-house history of the MAP, as well as a few other works, stress the humanitarian aspects of the Green Revolution.[135] Although humanitarianism alone is insufficient to explain the RF's motivations for establishing the MAP, many in Mexico gladly accepted the RF's aid. The early years of this cross-cultural encounter are best seen as a period in which, in the modernization of agriculture, the interests of the two cultures merged.

From the U.S. perspective, World War II made improved relations with Mexico, and increased output from its farms, essential. From the Mexican standpoint, the MAP addressed the desires of many parties who were interested in agriculture. Political leaders, the majority of the agricultural science community, many members of the general public, and numerous farmers (including land-reform beneficiaries) recognized that the agrarian reform had not solved the problems of rural Mexico, and that previous attempts to provide Mexican farmers with technical support had failed. The process of transformation started by the Mexican Revolution demanded the application of experimental science to modernize agriculture and thus, it was thought, bring prosperity to the countryside.

Many groups in Mexico benefited from the early years of the MAP. The project commenced at a crucial point of transition for the Mexican state. Having previously addressed some of the revolution's political goals for agriculture, such as attacking the hacienda system, the state now needed to focus on another goal: making rural Mexico productive and prosperous. Although this objective was not met in many areas of the country, the MAP aided the state in modernizing at least some segments of Mexican agriculture. The results of the project's research also helped the ruling party to diffuse attacks on its previous pro-agrarian policies, contributing to the government's ability to retain the agrarian legacy of Emiliano Zapata without having to pay the price of constant food shortages, which would certainly have occurred if levels of agricultural productivity remained what they were in the 1930s or if they had increased at a much slower rate than actually occurred. The MAP's work bolstered the state's image of competence among the general public, especially the more politically powerful and progressive Mexican farmers.[136] Even the land-reform sector benefited indirectly. Although peasant farmers often could not use the technologies developed by the MAP, it is likely that after 1950, without the gains in productivity generated in part by the Green Revolution, the question of land reform would have been more controversial, and the government might have been forced to terminate it much earlier than 1991.[137]

The MAP arrived at an opportune point in the professional evolution of the Mexican agricultural science community. It helped Mexico's agronomists rebuild their prestige and achieve the much-desired goal of creating a public image as scientists. The fellowship program for Mexican students exposed them to the cutting edge of U.S. science, providing Mexico with a sorely needed group of agronomists familiar with experimental methods. The MAP presented a model of farming that was advocated by many in Mexico at the time. Although the project's personnel neither appreciated the methods of peasant farming nor grasped some of the nuances of Mexican culture, they did provide a product that many wanted. Thus, the MAP was a program that met the needs of many in Mexico, at a crucial juncture in that country's history.

Notes

1. See Bruce H. Jennings, *Foundations of International Agricultural Research, Science and Politics in Mexican Agriculture* (Boulder, Colo.: Westview Press, 1988); Deborah Fitzgerald, "Exporting American Agriculture: The Rockefeller Foundation in Mexico, 1943–1953," Chapter 4 of the present volume; Cynthia Hewitt de Alcantara, *Modernizing Mexican Agriculture, Socioeconomic Implications of Technological Change* (Geneva: United Nations Research Institute for Social Development, 1976); Harry J. Cleaver, Jr., "The Origins of the Green Revolution" (Ph.D. dissertation, Stanford University, 1975).

2. Fitzgerald states that J. George Harrar noticed ambivalent attitudes toward the MAP in the Mexican agricultural science community. While this was true in situations in which MAP activities intruded directly on the projects of a given individual, such as Eduardo Limón, the director of the León, Guanajuato Experimentation Station, most in the scientific community supported the MAP, at least until 1950. See Fitzgerald, "Exporting."

3. For examples, see Alcantara, *Modernizing*; Angus Wright, *The Death of Ramón González: The Modern Agricultural Dilemma* (Austin: University of Texas Press, 1991), 171–73.

4. Pascual Gutiérrez Roldán, "Los Agrónomos y la Secretaría de Agricultura," *Agronómica* 1, 4 (January 1932): 6–9. Edmundo Flores, *Historias de Edmundo Flores, Autobiografía 1919–1950* (México, D.F.: Martín Casillas Editores, 1983), 98.

5. For the history of the ENA, see Ramón Fernández y Fernández, *Chapingo Hace 50 Años* (Chapingo: Colegio de Postgraduados, ENA, 1976). Regarding the lack of experimental training at the school, see Antonio C. Sandoval, "La ENA y el Consejo Nacional de Educación Superior y Investigación Científica," *Chapingo* 1, 2 (September 1935): 8–9.

6. During the last years of the Porfirio Díaz dictatorship and from 1915 to the early 1920s, much of the Ministry of Agriculture's effort was devoted to translating foreign agricultural bulletins. Mexican dependence on foreign agricultural science and technology prior to 1943 is discussed at length in Joseph Cotter, "Before the Green Revolution: Mexican Agricultural Science Policy 1920–1949," (Ph.D. dissertation, University of California at Santa Barbara, 1993). For scientific nationalism, see Eustacio L. Contreras, "Principios de una Organización, Sociedad Agronómica Nacional, Segundo Congreso Nacional Agronómico, 1922," Fondo Gonzalo Robles (hereafter FGR), Caja 5, Expediente 69, Archivo General de la Nación, México, D.F. (hereafter AGN); Rufino Monroy, "La Avicultura en México como se ha Visto y como Debe Verse," *Agros* 1, 1 (June 1923): 5; Ramón Fernández y Fernández, "Apuntes de las Clases de Entomología Dadas por el Prof. Alfonso Madriaga, el Segundo Semestre del Año de 1925, ENA," Caja 141, Document 212, Archivo de Ramón Fernández y Fernández, Zamora,

Michoacán, (hereafter ARFF); Gonzalo Robles, "Educación agrícola, propaganda," September 1921, FGR, Caja 49, Expediente 24, AGN.

7. Various works discuss this issue with respect to Latin American science in general. For example, see Francisco R. Sagasti, *Ciencia, Tecnología, y Desarrollo Latinoamericano* (México, D.F.: Fondo de Cultura Económica, 1981). For the Mexican agronomists, see Fitzgerald, "Exporting." This is not to imply that all Latin American science lacked an experimental tradition. For a contrary case, see Marcos Cueto, *Excelencia Científica en la Periferia, Actividades Científicas e Investigación Biomédica en el Perú 1890–1950* (Lima: Tarea, 1989).

8. For the agronomists and the revolution, see Fernández y Fernández, *Chapingo*, 39–59; Luís L. León, "La Actuación del Gremio Agronómico en la Reforma Agraria," *Boletín de la Sociedad Mexicana de Geografía y Estadística* 78 (1954): 57–71. For the agronomist's advocacy of land reform, see "Acta de la Inauguración de la ENA en Terrenos de la Hacienda de Chapingo, Mexico," Volume 168–2, Biblioteca del Colegio de Michoacán, Zamora, Michoacán (hereafter BCM).

9. For the agronomists' role in land reform, see Secretaría de Agricultura y Fomento, *Memoria de la Secretaría de Agricultura y Fomento Correspondiente al Periodo de lo de Agosto de 1924 al 31 de Julio de 1925* (Tacubaya, D.F.: 1927), 69–70; Leonardo Mendoza Vargas, "Como Proceder para que la Campaña de Abolición del Arado de Palo sea Eficiente," Segunda Reunión del Consejo Nacional Directivo de la Sociedad Agronómica Mexicana, April 14–19, 1941, Volume 186–5, BCM. In 1937 the government employed ten times more technical personnel to carry out the land reform than to work on research and in extension programs. See Secretaría de Hacienda y Crédito Público, *Presupuesto General de Egresos de la Federación para el Año de 1937* (México, D.F.: Secretaría de Hacienda y Crédito Público, 1937).

10. For example, during the 1930s agronomist Marte R. Gómez served as governor of Tamualipas, and agronomist Gilberto Fábila was the ruling Partido Nacional Revolucionario's Secretary of Agrarian Action.

11. Juan de Dios Bojórquez, "Momento Decisivo para el Gremio Agronómico," Trabajo Presentado al Segundo Congreso Nacional Agronómico, December 1, 1922, Volume 168–40, BCM. Carlos Terrazas Moro, "Agrónomos y Hacendados," *Agros* 1, 1 (June 1923): 2.

12. Eustacio L. Contreras, "Qué es un Agrónomo?" *Agros* 1, 2 (July 1923): 6. "El Próximo Congreso Agronómico," *El Nacional*, October 17, 1935. "La Cruzada de los Agrónomos," *Hoy*, April 26, 1941. Eyler N. Simpson used the term *"proyectismo"* to describe the Mexican government's programs of the 1930s, see *The Ejido, Mexico's Way Out* (Chapel Hill: University of North Carolina Press, 1937), 580.

13. Mexican research and extension programs during the 1920s and early 1930s are discussed at length in Cotter, "Before the Green Revolution." For examples of extension programs, see Secretaría de Agricultura y Fomento, *Memoria de la SAF*, 66–70; and Juan A. González to Agrónomos Regionales, March 10, 1926, FGR, Caja 5, Expediente 70, AGN. For research programs, see Enrique Beltrán, "La Dirección de Estudios Biológicos de la Secretaría de Fomento y el Instituto de Biología de la Universidad Nacional Autónoma," *Anales de la Sociedad Mexicana de Historia de la Ciencia y de la Tecnología* 1 (1969): 105–41; and Alejandro Brambila, "La Granja Experimental en Rodríguez, N. L.," *Irrigación en México* 1, 1 (May 1930): 44–48.

14. "El Cultivo del Maíz Debe Modernizarse," *El Universal*, December 17, 1933. "Campaña del maíz," *Extensión Agrícola* 1, 3 (April 1933): 115–16. Antonio Rivas Tagle, *El Cultivo Racional del Maíz* (Tacubaya, D.F.: Secretaría de Agricultura y Fomento, 1929); the last chapter describes U.S. work on corn hybridization.

15. These debates are discussed in Jesús Silva Herzog, *El Agrarismo Mexicano y la Reforma Agraria* (México, D.F.: Fondo de Cultura Económica, 1959).

16. Gutiérrez Roldán, "Los Agrónomos," 9; Bernardo Arrieta, "La Educación Agrícola como Fundamental de Reorganización de Nuestra Economía Rural," in Partido Nacional Revolucionario, *Los Problemas Agrícolas de México, Anales de la Economía Agrícola Mexicana* (México, D.F.: Partido Nacional Revolucionario, 1934), 255–57.

17. For calls for more research, see Secretaría de Agricultura y Fomento, "Sugestiones para la Elaboración del Plan Sexenal 1934–1939," mimeograph, 1933; and Emilio Alanis Patiño, *Diversos Aspectos de la Situación Agrícola de México* (México, D.F.: Instituto Mexicano de Estudios Agrícolas, 1934), 21, 36–39, 67–68. For scientific nationalism, see "Campos Experimentales para la Resolución del Problema Agrario," *El Nacional*, February 15, 1935; Gabriel Ítie, "Estaciones Agrícolas Experimentales," *Chapingo* 1, 2 (September 1935): 22; and Guillermo Gándara, "Desarrollo del Cultivo del Maíz en México," *Agricultura* 1, 3 (November–December 1937): 5–11. For examples of *"proyectismo,"* see Gutiérrez Roldán, "Los Agrónomos," 9; "La Agricultura y el Gobierno de 1934," *Germinal*, September 15, 1933.

18. Manuel Mesa A., "La Labor Oficial de los Agrónomos en México," *Agronómica* 1, 3 (December 1931): 34.

19. These reforms are discussed at length in Cotter, "Before the Green Revolution." For the founding of the *Instituto Biotécnico,* see Enrique Beltrán, "Instituto Biotécnico (1934–1940) de la Secretaría de Agricultura y Fomento," *Anales de la Sociedad Mexicana de Historia de la Ciencia y de la Tecnología* 1 (1969): 163–83.

20. See Secretaría de Agricultura y Fomento, *Memoria de Trabajos de las Direcciones de Agricultura y Ganadería Dependiente de la SAF durante el Periodo Presidencial de Lázaro Cárdenas* (México, D.F.: Departamento Autónomo de Publicaciones y Propaganda, 1940).

21. In 1935 the government announced that it was considering establishing an experimentation station in each of Mexico's 200 agricultural regions; see "Estaciones Agrícolas en Todo el País," *El Nacional*, November 29, 1935. The number of operating stations fluctuated annually, and scarce funds rendered some of them all but incapable of conducting research. For an example of the problems faced by research institutions, see Beltrán, "Instituto Biotécnico."

22. See Arnaldo Córdova, *La Política de Masas del Cardenismo* (México, D.F.: Ediciones Era, 1974), 106–13. He argues that the populist agenda of the Cárdenas government had a profound influence on the programs of the Secretaría de Agricultura y Fomento and other government agencies.

23. "Informe del General de División Lázaro Cárdenas ante el Congreso de la Unión, 1 de Septiembre de 1937," in *Palabras y Documentos Públicos de Lázaro Cárdenas, Informes del Gobierno y Mensajes Presidenciales de Año Nuevo 1929–1940* (México, D.F.: Siglo XXI Editores, 1978), 107. "Importación de Cereales Extranjeros," *El Nacional*, November 17, 1937. Gándara, "Desarrollo," 10. "Los Servícios de Extensión Agrícola en el País," *Agricultura* 1, 2 (September–October 1937): 2.

24. For example, see "Gran Escasez del Maíz en Aguascalientes," *El Universal*, February 2, 1937; Agustín García Rea and Isidoro Garibay Vázquez to Tranquilano Manríquez, January 21, 1937; and others, Caja AG-27bis-I, unnumbered Expediente, Archivo Histórico de Jalisco, Guadalajara, Jalisco (hereafter AHJ). Secretaría de Agricultura y Fomento, *Estudio Agro-económico del Maíz*, Tercera Parte (México, D.F.: Oficina de Publicaciones y Propaganda, 1940), 142.

25. For the campaign, see "Necesidad de Aumentar la Producción," *El Nacional*, January 6, 1938; and "Hay que Intensificar Nuestra Producción Agrícola," *El Universal*, June 17, 1938. By 1939 national average corn yields had increased 30 percent over 1937, a notably poor year, but national production still could not satisfy demand; see Secretaría de Agricultura y Fomento, *Memoria de la SAF de 10 de Septiembre de 1937 al 31 de Agosto de 1938* (México, D.F.: Departamento Autónomo de Publicaciones y Propaganda, 1938), 117.

26. For attacks on the agrarian reform, see Luis Medina, *Del Cardenismo al Avilacamachismo*. Vol. 18 of *Historia de la Revolución Mexicana, 1940–1952* (México, D.F.: El Colegio de México, 1978), 16–19. For attacks on the agronomists, see J. Rosario Heras Uriarte to Cárdenas, September 7, 1938; and "Estudio Formulado en la Oficina de Geografía Económico Agrícola por el Guillermo Rodríguez G. y Sometida a la Consideración del Presidente de la República con Fecha de 2 de Mayo de 1938," Lázaro Cárdenas del Río papers (hereafter Cárdenas papers), Expediente 437.1/556, AGN.

27. For an example, see Acuerdos Resolutivos de la Primera Convención Nacional de Productores de Oleaginosas," in Consejo Nacional de Agricultura, *Memoria de la Primera Convención Nacional de Productores de Oleaginosas* (México, D.F.: Consejo Nacional de Agricultura, 1939), 1–7.

28. Miguel García Cruz, "Ley de Servícios Agrícolas Nacionales," (no date), 10, Caja 142, Document 576, ARFF.

29. A report from the Ministry of Public Education argued that Mexican science needed help and would benefit from the influx of Spanish refugee scientists, but that since few of them were agricultural specialists, such specialists should be sought in the United States; see Enrique Arreguín, Jr., "Proyecto para la Organización de un Instituto Nacional de Investigaciones en Ciencias y Letras," September 1937, Cárdenas papers, Expediente 534/100, AGN. Some agronomists lamented the negative effect these crises had on the profession's prestige. For example, see Ramón Fernández y Fernández, "La Situación Actual de los Agrónomos," (no date), in Volume 168-40, BCM.

30. For example, see Secretaría de Agricultura y Fomento, *Memoria de la SAF de 10 de Septiembre de 1939 al 31 de Agosto de 1940* (México, D.F.: 1940), 103–15. During the late 1930s, the SAF spent a significant percentage of its research and extension budget on a new rural-education program, establishing primary and secondary agricultural schools, which represented a return to the methods of the 1920s. See Secretaría de Agricultura y Fomento, *Memoria de la SAF de 10 de Septiembre de 1938 al 31 de Agosto de 1939* (México, D.F.: 1939), 56–61.

31. See the collection of papers presented at the April 1941 meeting of the Sociedad Agronómica Mexicana that appear in Volume 186-5, BCM. In 1942 the organization's official plan to improve Mexican agriculture included "experimentation and application of genetic principals"; see Sociedad Agronómica Mexicana, *Plan para la Organización Agrícola de México* (México, D.F.: Sociedad Agronómica Mexicana, 1942), 3

32. Secretaría de Agricultura y Fomento, *Informe de Labores de la SAF del 10 de Septiembre de 1940 al 31 de Agosto de 1941* (México, D.F.: Editorial Cultura, 1941), 23. Secretaría de Agricultura y Fomento, *Informe de Labores de la SAF del 10 de Septiembre de 1941 al 31 de Agosto de 1942* (México, D.F.: Editorial Cultura, 1942), 37, 79.

33. Fitzgerald, "Exporting," 465. Elvin C. Stakman, Paul C. Mangelsdorf, and Richard Bradfield, "Agricultural Conditions and Problems in Mexico, Report of the Survey Commission of the Rockefeller Foundation," 1941, Rockefeller Foundation Archives (hereafter RFA), R.G. 1.1, Series 323, Box 5, Folder 37, Rockefeller Archive Center, North Tarrytown, N.Y. (hereafter RAC). Harry T. Edwards and James H. Kempton, "Report on the Agricultural Survey of Mexico for Complimentary Crops," September, 1941, R.G. 229, Box 95, National Archives Records Annex, Suitland, Md. (hereafter NARA).

34. For shortages of personnel trained in experimental methods, see Stakman, Manglesdorf, and Bradfield, "Agricultural Conditions," 52–56. "La Conferencia de Mesa Redonda en Yucatán," *Siembra* 3, 39–40 (June–July 1946): 34. For the lack of continuity in programs, see Instituto de Investigaciones Agrícolas, "Catalogo de las Cruzas Hechas para el Mejoramiento del Trigo. 1941 a la fecha" (no date), RFA, R.G. 6.13, Series 1.1, Box 27, Folder 299, RAC. For the influence of politics, see Lester D. Mallory to Cordell Hull, November 29, 1943, R.G. 166, Box 302, National Archives, Washington, D.C. (hereafter NA). M. M. Acosta to Javier Rojo Gómez, June 27, 1946, Manuel Avila Camacho papers (hereafter Avila Camacho papers), Expediente 506.1/30, AGN. For personalism see Stakman to Bradfield and Mangelsdorf, August 11, 1944, RFA, R.G. 1.1, Series 323, Box 1, Folder 8, RAC. For shortages of funding, see José María Aguirre and César Domínguez V. to Manuel Avila Camacho, November 7, 1941; Amado Oliveros P. to Avila Camacho, August 7, 1941, Avila Camacho papers, Expediente 506.17.14, AGN. A. R. Mann, "Observations in Mexico, Trip Report April 9–25, 1943," RFA, R.G. 1.1, Series 323, Box 1, Folder 4, RAC.

35. Secretaría de Agricultura y Fomento, *Informe de Labores de la SAF del 10 de Septiembre de 1945 al 31 de Agosto de 1946* (México, D.F.: 1946), 89, 397–98. For the Mexican government's

desire to produce more agricultural products for the war effort and to involve the agricultural sector in the industrialization of Mexico, see Secretaría de Agricultura y Fomento, *Plan de Movilización Agrícola* (México, D.F.: Secretaría de Agricultura y Fomento, 1942). For an example of farmer demands for technical support, see José Gutiérrez to Avila Camacho, October 29, 1942; and others, Avila Camacho papers, Expediente 485.1/10, AGN.

36. Mexican Agriculture, Annual Meeting, October 17, 1946, RFA, R.G. 6.13, Series 1.1, Box 32, Folder 359, RAC.

37. Roberto Gavaldón, "Algunos Datos sobre los Ingenieros Agrónomos," *Últimas Notícias, October* 14, 1941.

38. "El Viaje de los 5 Agrónomos es Descabellado y Muy Caro," *Últimas Notícias,* October 22, 1941. "Perifonemas," *Últimas Notícias,* June 18, 1942. Concha de Villareal, "La Sociedad Agronómica Mexicana solo ha Hecho Política en sus 25 Años de Existencia," *Excelsior,* May 28, 1945.

39. For the *Sinarquistas,* see Ignacio Gallegos and Bernardo Higareda L. to Avila Camacho, January 28, 1941, Avila Camacho papers, Expediente 133.2/6, AGN. Representatives of the Sindicato de Pequeñas Agricultores (Small Private Farmers Union) told Avila Camacho, "it is more valuable to listen to private farmers than to bureaucratic agronomists"; see Jose Díaz Moret to Avila Camacho, July 10, 1941, Avila Camacho papers, Expediente 506.1/7, AGN.

40. Benjamin Hinojosa, "Voz del Agora," *Últimas Notícias,* November 2, 1943. Edwin J. Wellhausen, Oral History, 40, 164–65, RFA, R.G. 13, RAC. Hesiquio Puga to Avila Camacho, April 14, 1942, Avila Camacho papers, Expediente 404.2/201, AGN. Norman Borlaug, Oral History, 155, RFA, R.G. 13, RAC.

41. Secretaría de Agricultura y Fomento, *Planeación Agrícola* (México, D.F.: Secretaría de Agricultura y Fomento, 1943), 1–5. "El Maíz. Cosecha de 1943. Sus Efectos en el Abastecimiento de México en Dicho Cereal para el Año en Curso. Medidas Adoptadas para Conjurar la Escasez," January 22, 1944, Caja 122, Document 72, ARFF.

42. Various farmer organizations, especially the cotton growers of La Laguna and Chihuahua, petitioned the president to exempt them from the program, and in one case offered to pay the cost of importing from the United States an amount of corn equal to what could be grown on 10 percent of their land. For example, see "La Siembra de Maíz en La Laguna no Conviene," *El Universal,* October 10, 1943.

43. J. Bernard Gibbs to Secretary of State, July 13, 1945, R.G. 166, Box 318, NA.

44. Edmundo Flores claims that Gómez was especially concerned with the agronomist's public image; see Flores, *Historias,* 396. For an example, see Marte R. Gómez "Es un Mito lo de Nuestro Fracaso Agrícola," *Siembra* 3, 42 (September 1946): 10–12, 30. In 1945, after the research projects of the MAP and the SAF had made some progress, Gómez recognized that making the results public would lead to feelings of "respect" for and "confidence" in the agronomists; see Gómez to Alfonso González Gallardo and others, September 24, 1945, RFA, R.G. 6.13, Series 1.1, Box 23, Folder 257, RAC.

45. For the anxious waiting, see George C. Payne to Raymond B. Fosdick, October 9, 1942, RFA, R.G. 1.2, Series 323, Box 10, Folder 63, RAC. According to Stakman, the SAF and RF personnel had very similar ideas regarding what the MAP should do; see Stakman to Manglesdorf and Bradfield, March 27, 1943, RFA, R.G. 1.1, Series 323, Box 1, Folder 4, RAC.

46. The RF thought that Harrar's appointment played a crucial role in generating Mexican support for the MAP. For example, see "Confidential Supplement to Report of E. C. Stakman on the MAP for the Period September 6 to October 7, 1944," RFA, R.G. 1.2, Series 323, Box 10, Folder 60, RAC.

47. During the late 1930s, the Mexican government sought assistance from the United States for pest-control campaigns for cotton and potatoes, fisheries research, and studies of livestock diseases, as did representatives of Mexican banana growers, sugarcane farmers, and poultry keepers. During the 1940s, the government obtained similar help for the lime and livestock industries, agricultural meteorology, and laboratory construction, and on some oc-

The Mexican Agricultural Project

casions Mexican farmers did so on their own. These cases are discussed in Cotter, "Before the Green Revolution."

48. For example, see "The Reminiscences of Elvin C. Stakman," interviews by Pauline Madow, Oral History Research Office, Columbia University, 1971, 1061, RFA, R.G. 13, RAC. Carl Tenbrook, "Report on Derringue, a Paralytic Disease of Cattle in Mexico," (no date), RFA, R.G. 6.13, Series 1.1, Box 14, Folder 164, RAC.

49. Fernández y Fernández, *Chapingo*, 111–16.

50. Sandoval, "La ENA," 8–9. Ramón Fernández y Fernández to Jefe, Departamento Técnico de Investigaciones Económicas, September 7, 1936, Box 131, Document 170. Marco Antonio Durán, "Al Margen de la Agrobiología," *Agricultura* 1, 8 (September–October 1938): 9–10. Emilio López Zamora, *La Situación del Distrito de Riego de El Mante* (México, D.F.: Editorial Ramírez Alenso, 1939), 74–75.

51. Marco Antonio Durán, "Experimentación Agrícola," Volume 186–5, BCM.

52. "Enseñanza Agrícola," SAM, Segunda Reunión del Consejo Nacional Directivo, Proyecto de Puntos Resolutivos, April 14–19, 1941, Volume 186-5, BCM. Durán, Fernández y Fernández, and Gilberto Fábila to Consejo Directivo de la ENA, August 25, 1945, Volume 168-2, BCM. Stakman to González Gallardo, June 20, 1945, RFA, R.G. 6.13, Series 1.1, Box 23, Folder 258, RAC. HHS Notes of Diary, October 13, 1945, RFA, R.G. 1.1, Series 323, Box 1, Folder 10, RAC.

53. Secretaría de Agricultura y Fomento, *Informe* (citation 35), 12–13,

54. Jennings, *Foundations*, 9–44; 45–118.

55. Some evidence suggests that the MAP's scientists did not seek to depoliticize Mexican agricultural science. See Stakman, Manglesdorf, and Bradfield, "Agricultural Conditions," 27

56. Ramón Fernández y Fernández, "La experimentación agrícola," *Germinal*, June 28, 1939. Marco Antonio Durán, *Apuntes de la Agrobiología*, (México, D.F.: Instituto Mexicano de Estudios Agrícolas, 1940), 7. Manuel Mesa A., "La Agricultura (Estudio para el Plan Sexenal)," January 12, 1939, 93, Box 153, Document 38, ARFF.

57. See the discussions in "Sociedad Agronómica Mexicana, Quinto Consejo Nacional Verificado en le Palacio de las Bellas Artes de la Ciudad de México durante las Días 15, 16, 17, y 18 de Mayo de 1945," Avila Camacho papers, Expediente 433/ 84, AGN.

58. "Por el Auge de la Producción, Discurso Presentado por el Ing. Gabriel Leyva V. a la Reunión del Cuarto Congreso Nacional de la Sociedad Agronómica Mexicana, México, D.F., el 10 de Septiembre de 1943," *Siembra* 1, 6 (September 1943): 1–3.

59. Although the factions in the conflict put forth these issues in the press, personalist and generational issues were—in the author's opinion—at least as important a cause as ideological considerations. In 1945 the faction led by agronomist Jesus Merino Fernández seized power in the SAM and sought to demonstrate the profession's pro-agrarianism and support for the peasantry by building closer ties with the CNC. The other faction backed the anti-agrarian position of candidate Alemán, arguing that land reform had accomplished its mission of destroying the hacienda system and that Mexico was entering a new era that demanded a revolution in production. For the conflict, see "Se Dividieron en dos Grupos los Agrónomos de la Capital," *El Universal*, May 12, 1945.

60. The SAM made this argument in 1942, stating that "no economic situation such as our own can be resolved without first seeing production," and "the only recourse that could permit the country to normalize its economy is the intensification of agricultural production." See Sociedad Agronómica Mexicana, Tercera Reunión del Consejo Nacional Directivo, "Nexos con Gobiernos Locales, con Miras a un Desenvovimiento de los Problemas Agrícolas Estatales," Volume 179-53, BCM.

61. Manuel Mesa A., *Agricultural Resources of Mexico* (Washington, D.C.: Inter-American Development Commission, 1946). Marco Antonio Durán, *Del Agrarismo a la Revolución Agrícola* (México, D.F.: Talleres Gráficos de la Nación, 1947), 11. Augusto Pérez Toro, "Mejoramiento de la Producción de Henequén," March 1942, Caja 101, Document 23, ARFF.

62. For the quotation, see Alcantara, *Modernizing*, 6. For the president's call to the agronomists, see Avila Camacho, *Address to the Mexican Agronomists* (México, D.F.: National and International Problems Series, Department of State for Foreign Affairs and the International Press Service Bureau, 1941).

63. "Puntos Ideológicos de la Sociedad Agronómica BAR que Fueron Aprobados en el Pleno Efectuado el Día 21 Agosto de 1941," Avila Camacho papers, Expediente 404.1/23, AGN.

64. Ejidatarios in Oaxaca, the peasant leagues of Puebla, Veracruz, and Durango, CNC leader Graciano Sánchez, and the National Private Farmers League (Sindicato de Pequeños Agricultores de la República) wrote the president to express their support; see Avila Camacho papers, Expediente 133.2/67, AGN. Graciano Sánchez to Avila Camacho, July 31, 1941, Avila Camacho papers, Expediente 133.2/6, AGN.

65. José Fernández Villegas to Avila Camacho, July 15, 1941, Avila Camacho papers, Expediente 133.2/67, AGN.

66. José Ch. Ramírez, "Problemas Nacionales: La Caña de Azucar," *Hoy*, November 22, 1941.

67. Three ENA students told the President, "The *pasantes* from the ENA warmly congratulate you in relation to the high ideas that you put forth yesterday about agrarian and agricultural policy, which we will make our own. But also, we protest energetically that at a banquet at the Club France, Misters Marte R. Gómez and César Mártino exhibit themselves as representatives of the nation's agronomists. An authentic Agronomic Society has prestigious technical specialists as representatives, not professional politicians that lack the majority of the profession's favor. We offer you our collaboration"; see José del Riego, Juan Francisco Kaldman, and Manuel Marcué P. to Avila Camacho, July 9, 1941, Avila Camacho papers, Expediente 133.2/67, AGN.

68. Salvador Lira López, Ramón Fernández y Fernández, and Dr. Quintín Olazcoaga thought that "North American technical experts could loan valuable cooperation" to programs to help the poorest of Mexico's rural poor; see *La Pobreza Rural en México* (México, D.F.: 1945), 109–10.

69. Harry M. Miller to Hanson, February 10, 1943, RFA, R.G. 1.1, Series 323, Box 1, Folder 4, RAC. In contrast, the Instituto Biotécnico's project list for 1939 contained fifty-one items, including studies related to: ecology; pest control; nutrition; the regulation of pesticides and fertilizers; the artificial insemination of livestock and bees; the promotion of edible-frog raising, particularly in indigenous communities; and many other diverse topics. This is an example of the *"proyectismo"* approach to agricultural science typical of the pre-1941 period. See Secretaría de Agricultura y Fomento, *Programa de Labores de la SAF 1939* (México, D.F.: SAF, 1939).

70. For example see "En Terrenos de la ENA se Efectuaron Pruebas del Nuevo Fumigante 'D-D,'" *Excelsior*, October 26, 1946.

71. Stakman described the IIA in a manner reminiscent of the agronomist's attacks on the SAF during the late 1930s; see "MAP, Report of Stakman for March 10 to April 3, 1948," RFA, R.G. 6.13, Series 1.1, Box 32, Folder 360, RAC.

72. "Progress Report, RF Agricultural Program in Mexico," November 1, 1944, RFA, R.G. 1.1, Series 323, Box 6, Folder 40, RAC. "Report of the Oficina de Estudios Especiales, 1 February 1943–1 June 1945," RFA, R.G. 1.1, Series 323, Box 6, Folder 38, RAC.

73. For a discussion of the gains in agricultural productivity during this period, see Reed Hertford, *Sources of Change in Mexican Agricultural Production 1940–1965* (Washington, D.C.: Economic Research Service, USDA, Foreign Agricultural Economic Report no. 73, 1971), 2. Some authors have downplayed the importance of the Green Revolution crop varieties in this process. For example, see Blanca Torres, *Historia de la Revolución Mexicana 1940–1952: Hacia la Utopia Industrial*, (México, D.F.: El Colegio de México, 1984), 82.

74. "MAP, Report of Stakman," 11–14. Excerpt of a letter, Weaver to Chester I. Barnard,

from Lima, Peru, re Columbia Agriculture Project, September 19, 1949, RFA, R.G. 1.1, Series 323, Box 3, Folder 19, RAC.

75. Annual Meeting of the Advisory Committee on Agricultural Activities, November 4–5, 1948, RFA, R.G. 6.13, Series 1.1, Box 32, Folder 360, RAC.

76. For example, the governor of Nuevo León was very enthusiastic about the new hybrid corn varieties and offered the SAF his full cooperation in getting the farmers of his state to use them; see Arturo B. de la Garza to Miguel Alemán, May 2, 1947, Miguel Alemán papers (hereafter Alemán papers), Expediente 506.1/18, AGN. For an example of the distribution of the seeds, see Roberto E. Palacios to Sánchez Colín, April 15, 1948, RFA, R.G. 6.13, Series 1.1, Box 14, Folder 167, RAC.

77. Alcantara, *Modernizing*, 37–42.

78. For example, see José Pérez Razo to Edmundo Taboada R., November 9, 1949, RFA, R.G. 6.13, Series 1.2, Box 68, Folder 767, RAC. Taboada to Harrar, June 4, 1948; Taboada to Harrar, June 5, 1948, RFA, R.G. 6.13, Series 1.1, Box 27, Folder 299, RAC.

79. Fitzgerald, "Exporting," 477–79.

80. At a meeting of the Advisory Committee on Inter-American Cooperation in Agricultural Education, RF official Harry M. Miller discussed "the problem of how to transmit technologies developed in the U.S. to the other American Republics"; see Advisory Committee on Inter-American Cooperation in Agricultural Education, January 25–27, 1945, R.G. 59, Box 469, Document 111.46/6–445, NA. For warnings, see comments by Professor Carl Sauer, University of California, on Vice President Wallace's idea that the foundation should work in Mexico in public health, nutrition, and agriculture, February 10, 1941, RFA, R.G. 1.1, Series 323, Box 1, Folder 2, RAC. Memo by Paul C. Manglesdorf, July 26, 1949, RFA, R.G. 1.1, Series 323, Box 3, Folder 18, RAC. For peasant methods, see Wellhausen, Oral History, 46–48, 66. Erlif V. Miller, John B. Pittner, Ricardo Villa J., and Carlos Romo G., "Population Density of Unirrigated Maize and Its Influence upon Fertilizer Efficiency in Central Mexico," (no date), RFA, R.G. 6.13, Series 1.1, Box 46, Folder 531, RAC.

81. One exception was French-trained agronomist Gabriel Ítie, who wrote, "Whatever system of cultivation, no matter how rudimentary, is the fruit of long observations by successive generations of laborers, it is a true adaptation to its medium that improves naturally, eliminates or modifies diverse and multiple adverse conditions, and arrives at a state of equilibrium that, if it does not lead to obtaining the highest yields or the best quality, does year after year provide the maximum coefficient of security in the results obtained"; see Ítie, "Estaciones," 2.

82. For an example, see División M. T. F., Dirección General de Agricultura, "Un Fuerte Azote para las Siembras de Frijol, Modo de Evitarlo," October 16, 1931, Box 90, Document 376, ARFF. The Green Revolution's technologies have been criticized for displacing more environmentally sound peasant methods. For example, see Wright, *The Death*, 171–87.

83. Rufino Monroy, "Cuál es la Raza de Gallinas que Más Conviene Explotar?" *Agros* 1, 5 (October 1923): 6.

84. This is not to say that RF personnel never considered addressing the needs of the small-scale peasant farmer. For example, see Stakman, "Report on Agricultural Activities in Mexico," February 3–May 20, 1943, 6, RFA, R.G. 1.2, Series 323, Box 10, Folder 60, RAC.

85. For example, see Durán, *Del Agrarismo*, 186–87.

86. William Vogt, "Confidential Memorandum Submitted to the Comisión Impulsora y Coordinadora de la Investigación Científica de México," November 1944, R.G. 59, Document File 1940–1944, Box 4032, unnumbered document, NA. He also presented these ideas in a monograph published in Spanish; see Guillermo Vogt, *El Hombre y la Tierra*, (México, D.F.: Secretaría de Educación Pública, Biblioteca Enciclopédica Popular, 1944).

87. A. R. Mann., "Topical Diary of Visit to MAP," September 12–October 6, 1946, RFA, R.G. 1.1, Series 323, Box 2, Folder 12, RAC. For Mexican advocacy of soil conservation, see David S. Ibarra, "Un Problema Nacional, la Erosión del Suelo y la Perdida de su Fertilidad,"

Segunda Reunión del Consejo Nacional Directivo de la Sociedad Agronómica Mexicana, April 14–19, 1941, Volume 186-5, BCM. Sociedad Agronómica Mexicana, "Investigación Agrícola," (no date), Caja 180, Document 53, ARFF.

88. From the beginning, some RF personnel thought that improved soil-management techniques should be an essential part of the MAP's program because of the high cost of chemical fertilizers in Mexico. For example, see Stakman, "Report on Agricultural Activities in Mexico," February 3–May 20, 1943, RFA, R.G. 1.2, Series 323, Box 10, Folder 60, RAC. In 1949, in response to Vogt's monograph, the MAP proposed a more-intensive campaign along these lines, but the Mexican government, which was committed to promoting chemical fertilizers, was not enthusiastic; see "Report of Harrar to the Natural Sciences Division and to the Advisory Committee on Agriculture," September 30, 1949–October 1, 1950, RFA, R.G. 1.2, Series 323, Box 10, Folder 61, RAC. Harrar to Lorenzo Patiño, January 19, 1950, RFA, R.G. 6.13, Series 1.1, Box 14, Folder 166, RAC.

89. J. I. Rodale to Agricultural Department, Government of Mexico, December 27, 1949; Sánchez Colín to Harrar, January 18, 1950, RFA, R.G. 6.13, Series 1.1, Box 15, Folder 170, RAC.

90. For example, see Alfonso E. Bravo, "La Agricultura Moderna," *Irrigación en México* 1, 3 (July 1930): 50–51; "Nuevas Especies en los Estados Unidos," *Revista Arrocera* 1, 1 (July 1935): 30; L. Iñíguez de la Torre, "El Rancho Americano y el Rancho Mexicano," *Jalisco Rural* 17, 7 (July 1937): 177–78; Zenon Lemanski, "Las Características del Servicio de la Organización de la Defensa de las Plantas en los Estados Unidos de Norte América," *Agricultura* 1, 5 (March–April 1938): 48–49.

91. One of the SAF's first research projects involved using a method developed by the German soil scientist Neubauer, in which the nutrient uptake of young plants is analyzed in order to discover the proper formula of chemical fertilizers required by a given soil; see Alejandro Brambila, Jr., *El Análisis Químico y la Fertilidad de los Suelos* (México, D.F.: Dirección General de Agricultura, 1928).

92. For an example of organic methods, see F. Jáquez, "Construcción de un Estercolero," *La Tierra* 1, 2 (May–June 1938): 12–13. Acknowledging the scarcity of dung, an SAF spokesman stated, "[this] clearly indicates the imperative necessity of employing other fertilizing agents, commonly known as commercial fertilizers or concentrated fertilizers, in a very ample scale"; see "La Escasez del Maíz en la República," *El Nacional*, February 26, 1938.

93. "Informe General y Presupuesto de la Oficina de Abonos Nacionales para los Años de 1938, 1939, y 1940," FGR, Caja 1, Expediente 8, AGN.

94. Pandurang Khankhoje, *Maíz Granada, "Zea Maíz Digitata" su Origen, Evolución, y Cultivo* (Tacubaya, D.F.: Talleres de la Oficina de Publicaciones y Propaganda, 1936).

95. Secretaría, *Informe*, 30, 55; Secretaría, *Informe*, 300–22 (citation 32).

96. Established as part of Mexico's wartime economic program, the firm applied for equipment purchases in the United States, saying, "It is well known that the only way to increase the agricultural production of the Mexican soil and to obtain its maximum yield, is to treat it conveniently with adequate and economical fertilizers, which at the present time are not available"; see "Application for Authority to Begin Construction Project to Order 41-1, by Manuel Noriega on behalf of Guanos y Fertilizantes de México, S.A.," (no date), R.G. 229, Box 592, NARA.

97. For example, see Miguel Brambila, "Memorandum al Sr. Ing. Gonzalo Robles, Referente a la Naturaleza de los Suelos de México y a su Necesidad de Fertilizantes," January 6, 1945, FGR, Caja 1, Expediente 6, AGN. Gómez cited the government's campaign to promote chemical-fertilizer use as an example of the SAF's progress toward modernizing Mexican agriculture; see Gómez, "Es un mito," 11–12.

98. For the surplus, see Norberto Aguirre, "Memorandum Sobre el Problema del Trigo," October 20, 1949, Alemán papers, Expediente 506.1/5, AGN. For an example of the role of fertilizers in Alemán-era campaigns to modernize agriculture, see *La Resurección de la Tierra*, Cuaderno de Cultura Popular, no. 8 (México, D.F.: Secretaría de Educación Pública, no date).

At times, representatives of the fertilizer manufacturer participated in the SAF's extension campaigns; see "Lista General del Personal que Actualmente se Encuentra Trabajando dentro del Programa Agrícola de Emergencia para Siembras de Maíz, Frijol, y Trigo, Planeado por la SAF, (no date), RFA, R.G. 6.13, Series 1.1, Box 42, Folder 478, RAC.

99. "El Problema Agrícola Nacional III," *Siembra* 3, 38 (May 1946): 6–7. "El Campesinado Nacional en las Conferencias de Mesa Redonda, en Aguascalientes," *Siembra* 3, 41 (August 1946): 19–20. Sociedades Locales de Crédito Ejidal, Tuxpám, Veracruz to Alemán, January 12, 1947, Alemán papers, Expediente 565.4/97, AGN. Comunidad Agraria Ejidal de Chupio to Alemán, April 7, 1947, Alemán papers, Expediente 506.1/13, AGN.

100. According to the directors of the CNC, "We commit the sin of utopians in asking that each peasant would select the seeds that he is going to sew"; see "El problema" (citation 99).

101. Torres, *Historia de la Revolución*, 58.

102. For an example, see "Formas de Cultivo que México Debe Adoptar en su Provecho," *El Nacional*, March 30, 1942.

103. For example, the campaign to promote the cultivation of castor beans under contract with the USDA turned into a disaster for many farmers, who were convinced by the SAF's literature that yields would be much higher than those actually obtained, which made the guaranteed price too low to cover production costs. See "No Cultivaron ya Higuerilla," *El Universal*, March 6, 1944.

104. U.S. Ambassador George C. Messersmith noticed this situation, and he dissuaded Nelson Rockefeller's Office of Inter-American Affairs from undertaking a program in Mexico to improve corn cultivation similar to programs the agency had established in various Latin American countries during the war; see Lester D. Mallory to Messersmith, August 16, 1944; Messersmith to Secretary of State, August 18, 1944, R.G. 166, Box 319, NA.

105. For example, the RF's policy with regard to publicity during its yellow-fever eradication campaign helped it attain its objective of helping the postrevolutionary regime of Alvaro Obregón build legitimacy and popular support during the early 1920s; see Armando Solórzano, "The Rockefeller Foundation in Revolutionary Mexico: Yellow Fever in Yucatan and Veracruz," chapter 3 of the present volume.

106. For an example of anti-U.S. sentiments, see Flores, *Historias*, 220. In 1940 the Liga de Agrónomos Socialistas expressed concern over the expansion of U.S. "imperialism" in Mexico as a result of the war, but many of its members, such as Ramón Fernández y Fernández, supported the MAP; see Comité Ejecutivo a Miembros de la Liga de Agrónomos Socialistas, October 28, 1940, Caja 180, Document 408, ARFF. Although some of the Mexican agronomists were fairly anti-American, this did not extend to the RF. An RF officer with much experience in Mexico told the RF's president, "the Foundation's relationship with Mexico was not strained even when the country had acute issues with the United States and its corporations"; see "Memorandum of John A. Ferrel regarding Mexico," February 3, 1941, RFA, R.G. 1.1, Series 323, Box 1, Folder 2, RAC.

107. For example, in discussing the activities of a U.S. plant geneticist who was doing research in Venezuela for the government and used the local press to generate much publicity concerning his project, an RF officer wrote, "Whether or not the same method would be as effective as one involving Americans occupying a less conspicuous role and giving the headlines as far as practicable to Mexicans may be doubted"; see "Notes of John A. Ferrel on Mexico," April 21–May 7, 1943, RFA, R.G. 1.1, Series 323, Box 1, Folder 5, RAC.

108. For example, in a 1946 speech to a meeting of the CNC, Gómez spoke proudly of the SAF's accomplishments in solving the problems of Mexican agriculture, and did not mention the RF or the MAP; see Gómez, "Es un mito," 10–12, 30.

109. For concern regarding the SAF's overoptimism, see Mann to H. vW., October 29, 1946, RFA, R.G. 1.1, Series 323, Box 2, Folder 13, RAC. When Watson Davis of the Science Service came to Mexico to visit the MAP, Harrar refused to give him any detailed information on the project and referred him instead to the SAF; see Harrar to Hanson, July 22, 1944, RFA,

R.G. 1.1, Series 323, Box 1, Folder 7, RAC. Harrar even refused a request for information about the MAP from the U.S. embassy in Mexico, which wanted to use the MAP as an example of the positive works that were a result of the Good Neighbor policy; see Harrar to Frank Blair Hanson, "Publicity in Mexico," June 22, 1943, RFA, R.G. 1.1, Series 323, Box 1, Folder 5, RAC.

110. Harrar to Warren Weaver, December 20, 1948, RFA, R.G. 1.2, Series 323, Box 11, Folder 64, RAC. Harrar to Weaver, November 29, 1949, RFA, R.G. 1.2, Series 323, Box 3, Folder 19, RAC. "Blueprint for Hungry Nations," *New York Times*, January 1, 1950.

111. Stakman, "Latin American Agricultural Institutions, Preliminary Report of a Trip," May 8–July 14, 1947, RFA, R.G. 6.13, Series 1.1, Box 32, Folder 360, RAC.

112. "MAP, Report of Stakman for March 10 to April 3, 1948," RFA, R.G. 1.1, Series 323, Box 3, Folder 19, RAC.

113. For discussions of patron-clientelism in Mexico, see Merilee Serrill Grindle, *Bureaucrats, Politicians, and Peasants in Mexico: A Case Study in Public Policy* (Berkeley: University of California Press, 1977), 26–40. Paul Friedrich, *Agrarian Revolt in a Mexican Village* (Chicago: University of Chicago Press, 1970).

114. See citation 112.

115. Wellhausen, Oral History, 44.

116. Lorenzo Martínez M. to Harrar, January 9, 1946; Eulogio Flores Aguirre to Harrar, May 6, 1946; Rodrigo Agostini V. and others to Harrar, July 2, 1947, RFA, R.G. 6.13, Series 1.1, Box 70, Folder 786, RAC.

117. Ramón Fernández y Fernández, "Nota Bibliográfica: *Informe de Labores de la SAF del 10 de Septiembre de 1943 al 31 de Agosto de 1944*," Caja 153, Document 109, ARFF.

118. Durán, *Del Agrarismo*, 180–81.

119. Manuel Mesa A., "La Agricultura en México, Informe Preparado para la Comisión Económica para América Latina," March 1950, FGR, Caja 11, Expediente 96, AGN.

120. Brillante Homenaje de la SAM," *El Universal*, March 20, 1944. Harrar to SAM, February 12, 1946, RFA, R.G. 6.13, Series 1.1, Box 3, Folder 36, RAC. "Mejoramiento de las Plantas y de las Animales," (no date), RFA, R.G. 6.13, Series 1.1, Folder 35, Box 395, RAC.

121. "Informe Rendido por el Comité Central Ejecutivo de la SAM a la Tercera Asamblea del Consejo Nacional Directivo," (no date), Caja 180, Document 82, ARFF. Arrieta to Weaver, November 6, 1946, RFA, R.G. 1.1, Series 323, Box 2, Folder 13, RAC. Antonio Marino A. to Harrar, August 2, 1945, RFA, R.G. 6.13, Series 1.1, Box 69, Folder 784, RAC.

122. "Puntos" (citation 63).

123. For example, in 1941 the SAM resolved "That Mexico should now present its agrarian and agricultural experience before the analysis of other nations of the Americas, so that we may exchange knowledge with them." Segunda Reunión del Consejo Nacional Directivo de la Sociedad Agronómica Mexicana, April 14–19, 1941, Volume 186-5, BCM.

124. Wellhausen, Oral History, 44.

125. At the 1945 meeting of the SAM, the agronomists in attendance devoted almost none of their discussions to matters related to experimental science; see SAM Quinto Consejo Nacional verificado en el Palacio de las Bellas Artes de la Ciudad de México durante los días 15, 16, 17, y 18 de Mayo de 1945, Avila Camacho papers, Expediente 433/84, AGN. Despite their pronouncements in favor of improving the profession's knowledge of experimental science, some agronomists still disdained the hands-on work it required, and preferred to occupy themselves in other activities; see Arrieta to Harrar, August 29, 1946, RFA, R.G. 6.13, Series 1.1, Box 69, Folder 785, RAC.

126. The only editorial I could locate that was somewhat hostile to the project was "Con Ayuda de Tres Sabios Americanos Resucitará la Agricultura," *Carta Seminal*, August 2, 1941. In contrast, *Últimas Notícias* reviewed the arrival of the site-survey team favorably; see Payne to Hanson, August 23, 1941, RFA, R.G. 1.1, Series 323, Box 11, Folder 72, RAC. For an example of a favorable review, see "La Fundación Rockefeller Da Importante Ayuda a Nuestra Agricultura," *La Prensa*, March 15, 1946.

127. Excerpt, Harrar to W. W., April 22, 1949, RFA, R.G. 1.1, Series 323, Box 3, Folder 17, RAC.

128. For example, the president of the National Confederation of Private Smallholders told Harrar, "The Institute that you dignifiedly represent has distinguished itself by its profound labor to improve, in the continent, the most essential dietary products," and he invited Harrar to the next meeting of his organization's National Directive Council; see Alfonso Castillo to Harrar, May 24, 1946, RFA, R.G. 6.13, Series 1.1, Box 20, Folder 216, RAC.

129. Comisión Ejidal of Mamaleón, Tamualipas to Alemán, December 20, 1948, Alemán papers, Expediente 506.1/50; Macario Barrios and others to Alemán, May 27, 1948, Alemán papers, Expediente 506.1/46; Gabriél Ramos Millán to Alemán, April 15, 1947, Alemán papers, Expediente 110.1/2, AGN.

130. Francisco Mejia and others to Alemán, April 24, 1949, Alemán papers, Expediente 565.4/97, AGN.

131. Richard Bradfield, "Report on a Trip to Mexico, 14 August–5 September 1953; 17–24 October 1953," RFA, R.G. 1.2, Series 323, Box 10, Folder 61, RAC. Wellhausen, Oral History, 174–81.

132. For example, E. Mendoza Vargas, "La Enseñanza y Extensión Agrícola, Ateneo Nacional Agronómico, Seminario Organizado con Motivo del Centenario de la ENA," February 1954, Volume 145-42, BCM. In 1976 Fernández y Fernández wrote, "The Rockefeller Foundation had to come in 1940 to put it [agricultural research] on a sound foundation. But, in agreement with ideas now in vogue, the acknowledgment of this fact, and even more any expression of thanks, is viewed poorly"; see *Chapingo*, 136.

133. Roberto Osoyo, "Mexico: From Food Deficits to Sufficiency," in *Symposium on the Strategy for the Conquest of Hunger* (New York: Rockefeller University, 1968), 10. In 1969 the Mexican government announced that it would issue a coin commemorating the MAP; see "Coin Column," *New York Times*, October 19, 1969.

134. Alcantara, *Modernizing*; Fitzgerald, "Exporting"; Wright, *The Death*; Steve Weismann, "Why the Population Bomb Is a Rockefeller Baby," *Ramparts* 18, 11 (May 1970): 43–48; Frances Moore Lappé and Joseph Collins, *Food First: Beyond the Myth of Scarcity* (New York: Ballantine Books, 1977); Gustavo Esteva, *The Struggle for Rural Mexico*, (South Hadley, Mass.: Bergin & Garvey Publishers, 1983); Andrew Pearse, *Seeds of Plenty, Seeds of Want* (Geneva: United Nations Research Institute for Social Development, 1979).

135. Elvin C. Stakman, Paul C. Manglesdorf, and Richard Bradfield, *Campaigns Against Hunger* (Cambridge: Belknap Press of Harvard University Press, 1967); Lester R. Brown, *Seeds of Change: The Green Revolution and Development in the 1970s* (New York: Praeger Publishers for the Overseas Development Council, 1970); Stanley R. Johnson, *The Green Revolution* (London: Hamish Hamilton, 1972).

136. Fitzgerald, "Exporting," 467.

137. President Carlos Salinas de Gortari proposed this on November 7, 1991; see "Mexico Seeks Land Reform, Bigger Farms," *Los Angeles Times*, November 8, 1991.

6

The Rockefeller Foundation's Medical Policy and Scientific Research in Latin America

The Case of Physiology

Marcos Cueto

D URING THE PAST few years, a number of studies have explored the role played by the Rockefeller Foundation in the organization of scientific work. They have considered some of the limitations of the RF's philanthropic policies abroad. In molecular biology, for example, the RF misconstrued important questions by its explicit application of techniques and models from the physical sciences to biological fields.[1] In Mexico, the RF's program of agricultural assistance failed to adapt to the local conditions of rural families, whose farms were small and labor-intensive.[2] In China, the RF's work in a single center of research and training showed little attention to how to diffuse the modern medical and educational resources so needed by the rest of the country.[3] The foundation's overseas programs tended to emphasize safe donations to scientists and institutions that were already well established and that shared the ideals of U.S. universities and laboratories.

This new literature has begun to effect a reconceptualization of the interweaving of science and philanthropy in many different countries and cultures. This chapter attempts to add to this historiography by examining in detail the problems found in the implementation of the RF's policies regarding physiological research in four Latin American countries.[4] In Latin America, the RF had an early interest in public health and education. Among all the philanthropic institutions supporting Latin American medical science, the Rockefeller Foundation was the oldest and, for much of the twentieth century, the most important. Latin American research in physiology developed significantly during the 1940s and 1950s, and researchers maintained an active relationship with the RF. The best-known indicator of this development was the Nobel Prize awarded to the Argentine Bernardo A. Houssay in 1947, for his researches on the glandular basis of sugar metabolism.[5]

The Medical Policy of the Rockefeller Foundation

During the years 1913 to 1940, the RF's medical activities in Latin America were directed by Americans and were concentrated in public health and the control of epidemics. In a second phase, between 1940 and the early 1960s, the foundation altered its emphasis to aid scientific education, individual researchers, and recognized institutions. The policy of awarding fellowships to study in American universities was common to both periods.

Activities during the first period were directed by the RF's International Health Commission (later renamed Board, and then Division), created for the purpose of extending the work of the Rockefeller Sanitary Commission for Eradication of Hookworm Disease to countries other than the United States. The RF sent U.S. physicians, scientists, and sanitary engineers to direct and reorganize public health boards and to direct campaigns to control infectious diseases, especially yellow fever and malaria. Their activities were inspired by the romantic goal of eradicating at least some endemic infectious diseases from the world.

These RF activities helped stabilize world trade and served the changing role of the United States in the world economy.[6] After the First World War, imports from the United States began to strain Latin America's economic links with Great Britain, which had been the region's main European commercial interest during the nineteenth century.[7] North American capital flowed into the larger Latin American countries, helping to consolidate an export economy based on raw materials. This trade in turn depended a great deal on the salubriousness of a country's ports and coastal regions.[8] Between 1910 and 1930, the U.S. government was concerned with protecting the health of U.S. investments and properties in those areas of Latin America deemed to be under their influence. Table 1 lists the amounts donated by the RF to countries of the region between 1913 and 1940. In those years, the RF spent $4 million in Brazil for the sole purpose of combatting yellow fever. In 1930 the RF gave over $1 million to equip a modern school of medicine in Sao Paulo, which came to be known as the "Rockefeller School."[9] For many years, Sao Paulo was able to maintain its leading position as the best medical school in Latin America.[10]

During the period from 1913 to 1940, the RF extended its interest in financing physiological research to Peru and Argentina (see Table 2). In Argentina, Bernardo A. Houssay directed the Physiology Institute of the University of Buenos Aires, where he studied the pituitary gland and pancreas, kidney function, and blood pressure. For many years, Argentina led all Latin American

Table 1. Donations (in US $) from the International Health Division of
the Rockefeller Foundation in Six Latin American Countries, 1913–1940

Country	Grants	Disease Control*	Public Health Services	Public Health Education	Totals
Argentina	2,418	91,459	15,000	—	108,877
Brazil	204,186	5,714,321	307,813	413,393	6,639,713
Chile	2,335	—	—	—	2,335
Colombia	40,745	410,695	66,400	940	518,780
Mexico	112,112	513,167	145,942	1,385	772,606
Peru	4,859	131,941	—	—	136,800

*Includes control and research of hookworm, malaria, respiratory diseases, and yellow fever,
and the collection and examination of animals.
Source: International Health Division, Appropriations for Work in Latin America, July 13, 1913
to December 11, 1940, RFA, R.G. 1.2, Series 300, Box 2, Folder 14, RAC.

countries in the number and quality of its investigators in physiology. In Peru,
experimental physiology was advanced by Carlos Monge Medrano and Alberto
Hurtado, the directors of the Institute of Andean Biology. This institute formed
part of the medical school of the University of San Marcos and was chiefly
devoted to the study of the effects of high altitude on the human body.[11] Argentine and Peruvian physiological achievements began to force a new appre-
ciation of Latin American research from U.S. scientists. As one U.S. physiologist declared in 1937, "Until I knew [Hurtado] I never gave South Americans
very much credit for anything."[12]

It was not until 1940, however, that the RF began to support biomedical
research in other Latin American countries. This support signified an expansion of the RF's medical program, which continued to be active in public
health. This new interest derived from several factors. The Second World War
interrupted the scientific relationship between the United States and Europe,
where the RF had carried out major programs.[13] This allowed the foundation
to pay attention to other regions of the world. With the war, Latin American
students of medicine could no longer pursue graduate studies in Europe and
began to turn their attention to U.S. universities.

Moreover, U.S. physiologists directed their work toward the solution of
war problems. For example, David Bruce Dill of the Harvard Fatigue Laboratory, a research center linked to Peruvian high-altitude investigators since the
early 1930s, performed routine tests on the ability of soldiers to withstand conditions such as those found in the African desert or Arctic zones.[14]

Diversion from the basic sciences was a common factor in wartime physiological studies in England and the United States, and it worked to the advantage of Latin American physiology research. Scientific development in the re-

Table 2. Rockefeller Foundation Donations (in US $) for Physiological
Research in Four Latin American Countries, 1926–1967

Years	Argentina	Mexico	Brazil	Peru
1926–30	800	—	—	—
1931–35	2,000	—	—	4,000
1936–40	25,000	5,000	—	7,860
1941–45	62,801	9,950	—	22,875
1946–50	46,000	39,660	3,300	56,000
1951–55	15,435	57,050	29,000	90,575
1956–60	40,800	71,750	89,976	7,550
1961–67	10,054	18,500	5,650	58,050
Totals	202,890	201,910	127,926	246,910

Source: RFA, R.G. 1.1., 1.2, Series 302, 323, 305, 331, RAC.

gion was due not only to decreased international competition but also to improved internal conditions for scientific work. During and after the Second
World War, the more-advanced countries of Latin America experienced a drastic reduction of trade, and the consequential weakening of the export sector,
which produced what has been called an "import substitution industrialization."[15] In Mexico and Argentina, greater state and private funds were chan-
neled to local research. Because physiology was one of the oldest and best-es-
tablished disciplines in these countries, it was able to attract the greatest share
of these new sources of support.

Another factor that influenced the new approach of the RF to Latin American science was the Good Neighbor policy begun by the administration of
Franklin D. Roosevelt. This attempt to achieve closer relations with Latin America
developed during the mid-1930s as part of U.S. efforts to check Germany's cultural influence there. The large communities of German and Italian immigrants
in Chile, Argentina, and southern Brazil were viewed as potential foci for an
increase of Nazi and Fascist influences in South America.[16]

During and, more clearly, after World War II, the United States set the promotion of North American culture in Latin America as an important goal.[17]
This concern contributed to the creation in 1939 of a Division of Cultural Relations in the U.S. Department of State, which in its early years devoted particular attention to inter-American relations.[18]

These factors combined to produce a reformulation of the RF's policy toward Latin American biomedical science.[19] As the 1939 report of the advisory
committee of the RF's Division of Natural Sciences stated,

> Disturbed conditions in many European countries are not propitious
> for steady attention to scientific studies. . . . Several members of the

Committee suggest that certain centers in Latin America might be advantageously stimulated by relatively small grants. . . . South Americans who have looked to Europe for scientific training are turning towards this country for instructive experience and discipline"[20]

Increasingly attentive to Latin American science, the RF formed its Inter-Divisional Committee on Latin America in 1943. With representatives from each divisio.1 of the foundation, the committee contributed to the elaboration and coordination of overall foundation policy. Beginning in the 1940s, policies were directed to the reform of medical training and research.[21] According to foundation officials, ensuring public health over the long term required the production of high-quality medical personnel at the local level.

Walter B. Cannon, possibly the most important figure in U.S. physiology, was responsible for the training of several leaders of experimental physiology in Latin America; they went to Harvard on RF fellowships. They included: Efren del Pozo and José Joaquín Izquierdo from Mexico; Juan T. Lewis and Oscar Orias from Argentina; Franklin Augusto de Moura Campos from Brazil; Joaquín V. Luco and Fernando Huidobro from Chile; and Humberto Aste-Salazar from Peru.[22] Cannon had been well known in Latin America since 1929, when the Mexican Izquierdo translated his *Course on Laboratory Physiology*. A short time later, the RF distributed 400 copies of the work, free of charge, to Latin American university professors.[23]

Cannon's views were instrumental in redesigning the RF policies toward Latin American physiological research. In the early days, the RF's Division of Medical Education had not given funds to purchase research instruments needed by returning Latin American fellows. As a result, first-rate, trained students returned to their countries with few opportunities for continuing research. Cannon complained vigorously to the RF of the futility of "leaving [men] without the facilities for investigation towards which their experience had been directed."[24] Eventually, the RF agreed to give such assistance and made it a regular practice.[25]

During the 1940s, the foundation recognized the limitations of the original program of training Latin American researchers, many of whom wasted their skills in an unfavorable milieu on returning home. Some of the first generation of physiologists who went to the United States for graduate education returned to their countries only to become propagandists of physiology or administrators of laboratories.

After the 1940s, medical education became one the RF's preoccupations in Latin America.[26] Following its rule of giving aid only where local conditions were known, the RF began to report on medical schools in the most important countries of the region.[27] These reports usually took a negative view of local medical education and research. According to the RF, the evils of Latin American medical schools included an excessive number of students in relation to teaching facilities, part-time teaching positions, poor development of the basic sciences, and a lack of laboratory work. All these characteristics were seen as a result of backwardness and the influence of the French system of medicine.

The RF believed that a profound reform was needed in Latin America, similar to the one that occurred in the United States, as embodied in Abraham Flexner's 1910 report.[28] This report recommended the closing of most U.S. medical schools, the raising of teaching standards in the preclinical sciences, and the use of Johns Hopkins as the prototype for medical education. The RF identified the educational and research model of the elite U.S. universities as the universal ideal.

In fact, a few countries could be considered as having an "excessive" number of medical schools. Argentina had five schools, only one of which functioned in Buenos Aires. Brazil, the largest nation in South America, had twelve schools, half of which were located in the two main cities, Sao Paulo (two) and Rio de Janeiro (four). Mexico had nine medical schools, with three operating in Mexico City. Finally, for many years Peru maintained only one medical school, in the capital city of Lima. The rest of the Latin American countries each had three or fewer medical schools.[29]

In most, experimental physiology was not an important part of the curriculum. Physicians held to the traditional idea that illness could be treated from clinical experience and observation. Their physiology courses usually consisted of oral repetition of French textbooks. The study of the function of organs was seen as a sterile exercise. However, a few pioneering scholars were able to bypass the system. Some of them, like Houssay, were born leaders who went through a process of self-education. Others managed to receive training in the United States or Europe.

The RF's activities in Latin America had three dimensions. First, by giving medical students scholarships to study in the United States, the foundation discouraged the Latin American physicians' preference for postgraduate training in Europe, especially in Paris. Second, because the RF had profound misgivings about traditional medical schools, it promoted the establishment of new educational and research centers independent of these institutions. Third, the RF gave money to outstanding researchers in the basic sciences. The donation of first-rate equipment to institutions willing to secure positions for full-time physiologists became one of the RF's main goals. Since physiology was one of the region's most developed research areas, it came to play an exemplary role in RF policy.

Donations to individual scientists were generally used for the purchase of research equipment and supplies and for travel abroad. The RF did not offer salaries because it believed that the problem of funding full time-positions had to be resolved at the local level. Encouraging local institutions to free scientists from the routine work of academic administration and medical practice was considered an important means to this end. In 1943, Sao Paulo's medical school was the only place in Latin America where full-time positions were officially and effectively established in all the preclinical departments. In Argentina, there were some full-time chairs, notably those of distinguished physiologists such as Houssay, but as a rule, Latin American professors of medicine worked on a part-time basis.[30]

Argentina and the Political Context

The application of the RF's medical policy was not free from tension and conflict. One difficulty was the political context in which Latin American scientists worked. The case of Bernardo A. Houssay is instructive. Beginning in the late 1920s, the Rockefeller Foundation supported Houssay's work and that of the Physiology Institute of the University of Buenos Aires. Of all the money awarded to Argentine physiology between 1929 and 1969, 67 percent was given to Houssay's institutes. During the decade from 1930 to 1940, Argentina was probably the most developed country in Latin America. It had been one of the largest exporters of wool, mutton, beef, and grain. A wealthy civilian elite maintained an oligarchic rule and achieved political stability and economic expansion.[31] Prosperity showed in the construction of railroads, the massive arrival of European immigrants, and the transformation of Buenos Aires into the so-called Paris of South America.[32] Houssay's institute had some of this cosmopolitan flavor and, beginning in the 1920s, was well known in foreign scientific circles. For example, when Harvard University reached its three hundredth anniversary in 1936 and invited leaders of science and scholarship to receive its honorary degrees at a commemorative ceremony, Houssay was among those invited. That year he also delivered the prestigious Dunham Lecture at Harvard.

The confidence the RF had in Houssay and his disciples, and the high degree of development that Argentine scientific life exhibited during the late 1930s, contributed to the establishment in 1941 of the Rio de la Plata and Andean Regional Office of the RF's International Health Division in Buenos Aires. The office was responsible for RF activities in Argentina, Bolivia, Chile, Ecuador, Paraguay, Peru, and Uruguay. The director of the Buenos Aires office, Lewis Wendell Hackett, came from the Rome office, which the RF had closed in 1940.

The relations between the foundation and Argentine science were significantly affected by the conflicts between Houssay and the U.S. government and the Argentine military regime during the middle 1940s and early 1950s. During those years, the cultural environment of Buenos Aires was affected by the emergence of nationalism and populism as major forces in Argentine politics. Standing out among the new political leaders were Juan Perón and a group of officers of the Argentine Army, who aspired to industrialize the country by rejecting the export economic model and severing ties with the United States.

The conflicts between Houssay and the new military government began in October 1943, when Houssay and other Argentine intellectuals signed a declaration "of effective democracy and American solidarity." They demanded the return of a constitutional government and the breaking of diplomatic relations with Rome and Berlin. The manifesto criticized the "Government of Colonels," with its strong pro-German sympathies, which had made Argentina the only Latin American country to declare neutrality. The Argentine government responded to the manifesto by summarily dismissing all the signers who held university and official positions, including Houssay.

Months later, in March 1944, Houssay and his disciples formed the Institute of Biology and Experimental Medicine in Buenos Aires, a private research center that received support from an Argentine philanthropist through the Juan Bautista Sauberán Foundation. This was the first establishment in Argentina of a privately supported center for scientific research, something that was rare in Latin America. The Sauberán Foundation initially contributed $21,000 to Houssay's center. The RF also supported the beginnings of this institute, giving $29,900 between 1944 and 1946.[33] A Committee on Houssay Journal Fund, formed by U.S. scientists including Herbert H. Evans, Walter B. Cannon, John F. Fulton, and Carl J. Wiggers, provided Houssay's institute with subscriptions to eighteen international journals for five years.[34]

From the mid-1940s through the mid-1950s, the new institute experienced many setbacks. Perón's supporters accused Houssay of practicing elitist teaching, of recruiting his disciples from the ranks of the oligarchy, and of receiving U.S. funds. In addition, Houssay was accused of favoring the Allies because he received financing from U.S. philanthropic institutions.

During the 1950s, Houssay's institute increased its international prestige, thanks to the Nobel Prize and the arrival of fourteen scientists from the United States, Canada, Europe, and Japan.[35] Twenty-two Latin Americans also worked there, among them eleven Brazilians and seven Chileans.[36] The editorial office of the *Acta Physiologica Latinoamericana*, a major scientific effort of the region, was located at the institute.[37] In 1957 and 1959, when Buenos Aires was chosen as the location for the Pan American Congress of Endocrinology and the International Physiology Congress, respectively, Houssay was named president of both events. During those years, his textbook *Human Physiology* was published in English by McGraw-Hill and received worldwide distribution. This was the first Latin American scientific work to be given such recognition.

Nevertheless, and in spite of the institute's prestige, problems developed between Houssay and the RF. They occurred because, in the foundation's opinion, the country's political situation made a program of medical reform difficult to realize. Robert Briggs Watson, the RF's regional director for South America, wrote from Rio de Janeiro:

> I entertain some doubts that further support should be given to Dr. Houssay, in spite of the fact he is a friend of ours and one of the world's best physiologists. This is for the reason that his Institute is a private affair, tolerated by a government which recognizes that the Houssay group is against it. Our policy has been to work through official agencies and rarely to help unofficial agencies.[38]

The answer to this letter was clear: "Feel free therefore to discourage any hope on the part of Dr. Houssay."[39] After 1950, the RF reduced its funding of Argentine physiology. In fact, during the first years of the Perón government, the RF did not support any important program in Argentina, with the exception of Houssay's institute.[40] In September 1949, the RF decided to close its regional office in Buenos Aires and to concentrate its efforts in Brazil instead. Houssay resented the decrease in support from the RF during the 1950s. In a letter to the foundation, Houssay pointed out that assistance to Argentina was "incomparably less than that given to Brazil, Chile, Uruguay and Colombia."[41] In 1959, an RF official reported that when he escorted Houssay during a visit to the United States, Houssay constantly asked him in front of other people, "Why has the Rockefeller Foundation forgotten me?"[42]

Other factors help to explain this decline in the assistance to Argentine research. In the late 1950s, Argentine physiology suffered from a lack of leadership. Some of Houssay's most distinguished disciples, such as Eduardo Braun Menendez, Oscar Orias, and Juan T. Lewis, had either died or retired from research. In 1959, Houssay was seventy-two years old; he was no longer active in research and there appeared to be no second in command. Houssay's new interest became the recently created Consejo Nacional de Investigaciones Científicas y Técnicas, for which he did receive substantial support from the RF. This support is partially explained by the RF's principle of always working with official agencies and its desire to institutionalize a stable environment for research. However, the new interest of the RF in a Latin American council of science and technology was ephemeral and did not signify the beginning of a new direction in its policy.

Mexico and the Search for New Institutions

While Argentina illustrated the political difficulties the RF experienced, Mexico was a clear example of the foundation's success in achieving one of its main objectives: the search for an institution, not associated with the traditional universities, that could serve as an example of achieving excellence by following U.S. standards. In Mexico, the RF encountered a rare combination, a medical center not associated with the main university and at which highly qualified researchers were employed. This was the National Institute of Cardiology. The institute's purpose was clinical practice rather than research in the basic sciences, but the broad definition of cardiology given by Ignacio Chavez, its dynamic director, along with the presence of Arturo Rosenblueth, opened up possibilities for research.[43]

Rosenblueth was a Mexican researcher who worked for fourteen years (1930–44) at the Harvard physiological laboratory of Walter B. Cannon. Rosenblueth assisted Cannon in verifying his theory of the chemical transmission of nervous impulses and was a crucial mediator for Spanish-speaking students at Harvard.[44] Following Rosenblueth's return to Mexico in 1944, Cannon told the RF, "He was the most energetic, the most stimulating and effective investigator I have had associated with me during four decades."[45]

During the early 1940s, Mexico experienced a new political situation that favored the activities of the U.S. philanthropic agency. The country began to abandon the populist trend that had characterized the years between the agrarian revolution of 1910 and the late 1930s. The climax of this trend had been reached in 1938 when President Lázaro Cárdenas expropriated North American and British oil companies. The next president, Avila Camacho, began the reversal of nationalist policies in 1940, and the next twenty years saw political stability, increased foreign investment, and closer relations with the U.S.[46] According to a report written by an officer of the RF, "Moving about in Mexico in 1941 one does not sense any hostility . . . to the capitalistic Yankee."[47] During those years, Mexican culture, maintaining its nationalism and pride in indigenous culture for local and international consumption, acquired a more cosmopolitan outlook.[48]

The government supported the National Institute of Cardiology through the health department and a subsidy that in 1945 amounted to approximately one million Mexican pesos, (U.S. $200,000).[49] The original facilities included an outpatient clinic, laboratories, and a hospital with 120 beds and a staff of 45 physicians, 14 of whom were full-time.[50] Rosenblueth headed the Department of Physiology and Pharmacology, which was a research unit with, in 1946, a budget of U.S. $18,690.[51]

In that year, the RF donated an additional $18,000 to the section directed by Rosenblueth. His research program was concerned principally with the physiology and pharmacology of heart muscle and its blood supply through the coronary arteries, together with certain basic studies of nerve conduction and reflex action. The latter topic was an extension of Rosenblueth's work during his Harvard years. Between 1949 and 1954, forty-eight papers dealing with these subjects were published by members of the Institute, approximately half of them in English.[52]

As in tne case of Buenos Aires, the RF had confidence in Rosenblueth's work. Except for Houssay's institute in Buenos Aires, which was "under the shadow of constant political threat," it was difficult, according to an RF official, to find another group in Latin America "more likely to forward the development of experimental medicine" than Rosenblueth's group at the National Institute of Cardiology.[53] From 1944 to January 1954, the Mexican institute received a total of $110,100 in RF donations. Some $60,950 was awarded to Rosenblueth's department. Another important grant was made between 1947 and 1952. In 1947 the RF donated $27,500 for a joint project in mathematical biology between the institute and the Massachusetts Institute of Technology.[54] An agreement was reached by which Norbert Weiner of MIT spent six months every other year in Mexico, and Rosenblueth spent six months of the intervening years at Cambridge. Rosenblueth and Weiner came to be considered the founders of the new science of cybernetics.[55]

During his Mexican years, Rosenblueth became increasingly involved with administrative duties, which limited his time for research. Breaking local tradition, institute director Chavez assigned him a salary higher than that of any other scientist in the country, but to safeguard Rosenblueth from public criticism, the arrangement was made confidential, with part of the salary charged to a temporary special grant.[56] The resulting insecurity obliged Rosenblueth to become a scientific laboratory administrator, and although he did not stop publishing in mainstream journals, he became a popularizer of research. In 1953 he was teaching advanced physiology, electricity, mathematics, and scientific method at the Colegio de México and the National Institute of Cardiology.[57] In 1960 he became the first director of the prestigious Center for Research and Advanced Studies of the National Polytechnic Institute. As Benton notes, Rosenblueth's Mexican career can be seen as the transformation of an international laboratory researcher into a national administrative scientist.[58]

Brazil and Full-Time Positions for Physiology

Besides the political and institutional problems, and mainly because most scientists were physicians who maintained private practices, the RF had difficulty in securing full-time positions for physiology. The Latin American medical professor was a practitioner first and only incidentally a teacher of medicine. Reviewing the case of one Brazilian scientist will illustrate the situation.

Between 1953 and 1959, the RF donated $36,350 to the Department of Physiology of the University of Bahia, one of the lesser-known schools in Brazil. This move was part of a policy to develop provincial centers that could compete with the traditional medical schools. In Bahia, Jorge Novis had studied the chemical composition and nutritive value of cocoa—of great regional interest, given the economic importance of the commodity for the state of Bahia. But when he became a full-time professor in 1954, Novis did not leave his private practice. He explained his refusal to dedicate himself to the university in a 1955 letter directed to an RF official:

> In a country with high inflation such as ours, in which the cost of living increases every month, I would be called irresponsible if I risked the shelter I owe to my family. . . . I must ask: will I be better able to produce more for my laboratory with an insufficient and problematic full-time position, or without full-time, working a minimum of 36 hours a week, teaching and doing research, as I now do? I respond affirmatively to the last hypothesis.[59]

Some years later, Jose Simoes, another physiology professor from the same department, repeated the argument in a letter directed to the foundation: "I wish the full time, but not a hungry salary!"[60]

Foundation officials disapproved of these arguments. In 1955, the head of the RF's Rio de Janeiro Regional Office wrote that Novis's salary demands were absurd for Brazil, and that Novis would always find an excuse for not accepting a full-time position.[61] However, some officials were aware of the difficulties in applying their policy. According to the RF's representative in Peru, although the efficiency of full-time service was generally locally recognized, "the fear of losing the position, in possible changes of government," contributed to the lack of progress in this area.[62]

The Bahia case demonstrates the resistance encountered by the RF's medical policy. But it was not representative of the relationship between the foundation and Brazilian medicine. In general, physiology received little support because it was the weakest of the biomedical sciences in Brazil.[63] But, the RF maintained a great interest in medicine and science in this country.[64] Starting in 1954, the RF undertook a program to develop selected departments—mainly in the basic sciences—to improve the quality of research and teaching. During the period from 1954 to 1958, it spent $1,518,900 in the development of hospitals, medical schools, and medical research institutes in Brazil.[65]

This interest partially followed from the fact that in 1954 Brazil was the only Latin American country that was attempting a reform of its medical schools. "Federalization" reduced the number of medical students, with the consequence that while in 1950 seven medical schools admitted 1,512 students, four years later, the same schools admitted only 761.[66]

As in Mexico in the 1940s, the RF found political conditions in Brazil in the 1950s favorable for the development of its policies. In 1954 the country departed from the populist model of independent capitalist development that President Getulio Vargas had initiated in 1930. The election of 1955 elevated Juscelino Kubitschek to the presidency, and he stressed the modernization of the country and the acceleration of economic growth. Foreign investors and local entrepreneurs transformed the Amazon country into an agrarian-industrial country with a base of heavy industry. The new capital of Brasilia, constructed at a rural site in central Brazil, 600 miles northwest of Rio de Janeiro, became the symbol of modernity.[67] RF officials and the Brazilian government shared two interests: a concern for the development of the provinces and a desire to achieve change in a short period of years.

However, the broader policy the RF developed in Brazil was not essentially different from the elitist practices it had used elsewhere in Latin America. According to Robert Briggs Watson, the official in charge of the RF's Brazilian activities, the objective of his program was to strengthen a small number of schools and, in turn, make possible the utilization of these points for the training of other Brazilians. "We have never," he told Warren Weaver, "interested ourselves in quantity production of physicians."[68]

Peruvian High-Altitude Research

For several reasons, Peruvian physiologists received sizeable grants for equipment (see Table 2). In the first place, Peru offered a unique strategic advantage for the study of human life at high altitudes. In Peru the Andes rise precipitously, reaching altitudes of 14,000 to 20,000 feet within 100 miles of the coast. In the 1950s, more than 4,000,000 Peruvians lived at elevations between 6,000 and 14,000 feet, and a few thousand lived in mining towns above 15,000 feet. This fact afforded researchers ample opportunity to observe how the human body reacted to low oxygen pressure. The Institute of Andean Biology established three laboratories: one at Lima, the sea-level capital of the country; a second at Huancayo, an agricultural town at 10,700 feet; and a third at Morococha, a mining town located at 14,900 feet.[69]

Alberto Hurtado was the champion of the institute's international patronage. The first equipment grant to Peruvian physiology was made in 1934, when Hurtado returned from his RF postdoctoral fellowship at the University of

Rochester. The foundation gave him the necessary equipment for respiratory investigation, chiefly in connection with pulmonary capacity, blood gases, routine morphological blood tests, and electrocardiography. With this help, Hurtado was able to equip a small laboratory in Lima. He was probably the first returning Latin American medical fellow of the RF to receive such a grant-in-aid.

From that time on, Hurtado was able to obtain funds when other Peruvian fellows returned from postgraduate training and needed special instruments. By 1954, ten members of the institute's staff had studied abroad on Rockefeller Fellowships. Fully three-fourths of the RF grants went to providing research equipment.[70] As an RF official observed in 1951, "In going over the various extensive files in this case I could not avoid the impression that Hurtado is a skilled and rather persistent asker."[71]

Hurtado and the other leader of Peruvian physiology, Carlos Monge Medrano, were appointed to different university and state positions during the various governments. For example, Hurtado was Dean of the School of Medicine and Minister of Public Health.[72] A final factor explaining the RF's donations can be found in the political situation of Peru between 1934 and 1960. During those years, conservative governments, mostly military regimes, maintained good relations with the U.S.[73]

But Peru never overcame the export-economy model, probably because its progress toward industrialization was slower.[74] There was a lack of private funds that could support local research. Therefore, Peruvian physiologists became more and more dependent on the donations of U.S. philanthropic agencies. Although the unique high-altitude laboratories were used by local researchers, they frequently had to be prepared for foreign investigators. In addition, high-altitude studies developed during the 1950s, thanks to the U.S. Air Force, which provided salaries for some full-time researchers. According to Hurtado, nothing could be done without this foreign help.[75]

An aspect of RF policy that may have had a negative effect on Peru was the technological dependence that its donations produced. This dependency consisted of emphasizing the donation of the most-sophisticated and expensive research equipment made in the United States. Likewise, most of the chemical substances donated were produced only in the United States, and the foundation was the principal agent for access to these products. Over the years the RF donated to the Peruvian scientists practically the same modern equipment that the Harvard Fatigue Laboratory contained. In 1955, a physiologist from the University of Rochester worked in Peru for some months. During the course of his stay, he became dissatisfied with the Institute of Andean Biology because most Peruvian researchers used techniques learned in the U.S. only for certain definite problems, from which they rarely deviated. Their technique was excellent, but there was no latitude for flexibility in their work, and as a result they became "supertechnicians instead of original workers."[76]

In this way, a false association was made between original research and possession of modern equipment. This was an important reason for the later decline in physiological research. When, in the early 1950s, the RF changed its priorities in Peru from biomedical to agricultural research, physiologists in most countries were unable to acquire the sort of equipment they normally used. The case of Peru depicts in an extreme way one of the main differences between the reform of research in the United States, and Latin American biomedical research during the 1950s: the almost complete absence of local university endowment.

Evaluating the Impact

The impact of the RF's medical policy in these four Latin American countries can best be understood as the consequence of two conflicting academic and cultural traditions. The contrast is illustrated in one of the first letters of Houssay to Cannon, in which he noted, "Je lis l'anglais mais je ne le parle pas."[77]

The Latin American tradition—influenced by French medical education—was characterized by remarkably uneven scientific development, medical schools with unwieldy educational infrastructures, and no conception of science as an exclusive activity. However, some able scholars could supplement their training through self-education and contact with foreign researchers. A few Latin American medical schools managed to develop a fragile balance, elitist but efficient, between a few first-class laboratories and chairs directed to provide massive professional training for medical students. This balance allowed some professors free time to work on experimental physiology.

Houssay carried out his research ideas through the brightest medical students, selected from a large class of physiology. Taking into account that there was no formal system of attracting young talent to science, Houssay needed to be in contact with "overcrowded" classes. The outstanding research produced by Latin American physiology during the 1940s was the result not of an environment ripe for academic professionalization but of a special combination of science and professional education, in addition to the sheer ability of some brilliant people.

The tradition that inspired the RF's policies was the U.S. academic model, characterized by a critical mass of researchers, convenient teaching and research facilities, and the professionalization of science. The same model was promoted in Latin America, where modern science had developed in only a few laboratories and where evidence suggests that there was a shortage of doctors in terms of the needs of the population.

Some local researchers stood out as able intermediaries in relations between the RF and Latin American science. They identified with foundation goals but were capable of weighing both the values of their communities and the modernizing policies encouraged from abroad. In so diffuse a setting, these scientists could maintain some autonomy and initiative, thereby regulating the pace and mode of change.

In their local environment, Latin Americans accepted compromises with the traditional forces that dominated their universities, and even declined opportunities to develop academic careers in the United States. In 1943, after being dismissed from the University of Buenos Aires, Houssay received several invitations to work in the United States. He declined them all.[78] Some years later, he was offered a position that included moving his research staff from Buenos Aires to Washington. In a handwritten letter, he explained that he did not accept the offer because his life was consecrated to "impossible" objectives, such as the scientific development of the country in which he "was born, educated, had friends," and in which he "fought, learned, taught."[79]

Among the second generation of the RF-supported Latin American physiologists, tension developed more clearly. Houssay's prestige and seniority could ease his life in an environment ambivalent to research. Younger researchers were trained to work in a developed scientific environment while living in an underdeveloped one. The tragedy of their success was that they had developed too rapidly within a university framework that had changed too little. The doubts these scientists experienced is reflected in a letter written by Juan T. Lewis, a former student of Cannon and Houssay's, in which he speculated what his life would have been if he had remained in the United States after concluding his graduate studies: "I have been very happy here, but I think I would have done more for science there."[80] The degree of identification with the RF's policies was mediated by recognition of the superiority of U.S. medical training and research. In 1937, Hurtado was elected a member of the Peruvian Academy of Medicine. The leaders of this traditional institution, dominated by French-trained physicians, rendered homage to his work as a "good demonstration of American training." Hurtado wrote to Robert A. Lambert of the RF, pointing out the significance of the ceremony, "I think that the day is not far when American medicine will have a decisive influence in our teaching and research. I will do my best to contribute to this end."[81]

The exaltation of U.S. medical education reveals one of the main assumptions underlying foundation policies: that all developing academic institutions should develop in the same manner and direction. Change in Latin America was understood to be the result of an external stimulus, in this case the RF.[82]

The desire to replicate in Latin America the conditions under which RF fellows were trained in the States appears in a letter of an Argentine physiologist trained at Harvard: "Our actual conditions . . . are similar to those existing at Harvard when President Eliot started his work."[83]

Physiologists probably were, as Benton has said of Rosenblueth, "caught in a boundary region of cultures."[84] To carry out first-class research according to the standards learned in the United States, they were increasingly compelled to break with the structures of university education in their countries. At the same time, there were few alternatives outside the system.

The model exported by the RF had a profound effect on the reorganization of local scientific communities. North American influence in Latin American biomedical research increased in proportion, or at least in parallel, to the decline of European influence. Student scholarships and travel grants to visit U.S. universities consolidated U.S. influence in Latin American medical education and research after the Second World War. The RF helped sustain this change by promoting an international network of scientists. In the case of physiology, two circuits existed. The first and most extensive comprised researchers in endocrinology and neurophysiology; they joined Houssay in Buenos Aires and Cannon in Boston. The second circuit developed among the Peruvian physiologists: Hugo Chiodi, who directed an Argentine provincial research center, the Institute of High Altitude Biology of Tucumán; William McCann of the University of Rochester; and David Bruce Dill of the Harvard Fatigue Laboratory. This group in this circuit was smaller and concentrated on the study of high-altitude and respiration physiology.

During the period from 1913 to 1940, the relations between Latin American scientists and the United States could be conceived as a triangle without a base: Latin Americans were more likely to have contact with U.S. scientists than with other Latin American researchers. For example, in the early 1940s, the RF wanted to make Buenos Aires a South American physiology training center, but these efforts failed because, as one official revealed, "the young man wants to come to the U.S."[85] The amount of intraregional communication increased beginning in 1940. However, the base of the triangle always remained diffuse.

With the help of the RF, Argentines were able to extend their influence to other South American countries during the 1950s. Houssay and his disciples were instrumental in developing physiology in Brazil. The Argentine institute loaned laboratory personnel, and Houssay himself spent considerable time in Porto Alegre to develop the Institute of Experimental Physiology of the University of Rio Grande do Sul.[86] In addition, the Argentine Miguel Covian, whom Houssay considered the best researcher he had, moved to the school of Ribeirao Preto, in Sao Paulo.[87] During the 1950s, the Mexicans were able to attract some foreigners to work in the laboratory of physiology at the National

Institute of Cardiology. However, in comparison with that of Argentina, their regional influence was always smaller and much more linked to clinical aspects of biomedical research.

The RF's medical policy obtained contradictory results when it concentrated in a few centers of excellence, such as the medical schools of the universities of del Valle in Cali, Colombia, the Católica in Santiago, Chile, and Ribeirao Preto in Sao Paulo, Brazil.[88] Despite the fact that full-time positions were implemented in these schools, the expected multiplying effect in the structure of biomedical research in these countries was not produced. On the contrary, this policy contributed to the beginning of an obstacle in the development of the Latin American university system: the opposition between a few centers of academic excellence that had little effect beyond the local elite, and the mass teaching centers where the lack of material resources hampered the development of high-level research.[89]

Physiological research experienced new problems because of the process of student overcrowding (in Spanish, *masificación*) of the Latin American university, which began around the 1950s and continues to the present. *Masificación* was the product of societies in which the middle class grew without having many channels for upward mobility and in which the political elite did not reformulate the functions of higher learning. More students with the same number of professors and resources meant the destruction of the fragile balance between scientific research and professional education achieved in the earlier years of the twentieth century.

Conclusions

The relationship between the RF and Latin American physiology reveals the inadequacies in transferring a model of scientific work from one culture to another, the rigidities of U.S. philanthropy, and the limitations of centers of local excellence in an overall context lacking in economic development and technological sophistication. In this sense, it resembles the RF experience in China.

However, there are some differences with the other cases mentioned at the beginning of the chapter. Physiology was one of the few areas in which Latin American investigators had developed a native capacity before the arrival of U.S. philanthropy. In Mexico and China, there were either just a few researchers or none in the areas in which the RF intervened. The study on China suggests that the RF had a pervasive effect in shaping the work of the few local researchers who were alienated from their cultural milieu. Likewise, some studies on the early RF public health activities in Latin America have associated U.S. philanthropy with local passivity and lack of initiative.[90]

The relationship between the RF and Latin American physiology was not a case of unilateral diffusion. Conflict and local resistance were important ingredients of the story. Houssay, Rosenblueth, and Hurtado were willing to "Americanize" themselves, but at the same time maintained their ties with the traditional structure of their medical schools.

The foundation supported Latin American physiology not only as a result of the degree of excellence attained in certain countries but also as a form of spreading an academic model inspired by the elite universities of the United States. RF officials hoped that by providing examples of original, dedicated work, they could incite other scientists to transform the scientific structure of their counties. Nevertheless, this did not occur. Foundation support did not produce the competence or the imitation of the most prominent cases of excellence. The failure of the modernization of Latin American biomedical science during the 1950s and 1960s was due to internal factors as well—beyond the action of the philanthropic agency and beyond the scope of this chapter.

The notions of replication and change carried out by external agents, which were the basis of RF policies, revealed at the end little regard for the social, political, and cultural conditions of scientific work in Latin America. They were also a misreading of the U.S. experience of medical reform of the late nineteenth century. In the past few years, several works of U.S. scholars have confirmed the idea that medical reform was a social process that involved many years and different people, not occurring solely in one year, 1910, or thanks to one person, Abraham Flexner.[91] The inability to understand science in the periphery in its own cultural terms and not as an example of tradition or modernity is a problem that remains to this day.

Notes

1. See Pnina G. Abir-Am, "The Discourse of Physical Power and Biological Knowledge in the 1930s: A Reappraisal of the Rockefeller's 'Policy' in Molecular Biology," *Social Studies of Science* 12 (1982): 341–82.

2. Deborah Fitzgerald, "Exporting American Agriculture: The Rockefeller Foundation in Mexico, 1943–1953," chapter 4 of the present volume.

3. Mary Brown Bullock, *An American Transplant: The Rockefeller Foundation and Peking Union Medical College* (Berkeley: University of California Press, 1980).

4. Very little is known about the worldwide diffusion of U.S. philanthropy. Works that go some way remedying this situation for the case of Latin America are: Fitzgerald, "Exporting American Agriculture"; Guillermo Arbona and Annete B. Ramirez de Arellano, *Regionalization of Health Services: The Puerto Rican Experience* (Oxford: Oxford University Press, 1978); Kenneth

F. Kiple, *The Caribbean Slave: A Biological History* (Cambridge: Cambridge University Press, 1984); Marcos Cueto, "Excellence in the Periphery: Scientific Activities and Biomedical Sciences in Peru" (Ph.D. diss., Columbia University, 1988).

5. For a biographical sketch of Houssay, see Ariel Barrios Medina, "Bernardo Alberto Houssay (1887–1971): Un Esbozo Biográfico," *Interciencia* 12 (1987): 290–99.

6. Two fascinating testimonies of the interweaving of U.S. interest and RF public health goals appear in the memoirs written by two of the RF physicians who worked in Peru and Brazil: Henry Hanson, *The Pied Piper of Peru, Dr. Henry Hanson's Fight against "Yellow Jack" and Bubonic Plague in South America 1919–1922*, ed. Doris M. Hurnie (Jacksonville: Convention Press, 1966); and Hugh H. Smith, *Life's a Pleasant Institution: The Peregrinations of a Rockefeller Doctor* (Tucson, Ariz.: Smith, 1978).

7. For a discussion of this problem, see Thomas E. Skidmore and Peter H. Smith, *Modern Latin America* (New York: Oxford University Press, 1984), 46–51.

8. The best, single historical volume of this period for the four countries studied in this chapter is, Roberto Cortes Conde and Stanley Stein, eds., *Latin America, A Guide to Economic History 1830–1930* (Berkeley: University of California Press, 1977).

9. University of Sao Paulo, Faculty of Medicine, Rockefeller Foundation Archives (hereafter RFA), R.G. 1.2, Series 305, Box 17, Folder 151, Rockefeller Archive Center (hereafter RAC).

10. This opinion is confirmed in a letter from Fasciolo to Miller, June 19, 1957, RFA, R.G. 1.2, Series 301, Box 5, Folder 34, RAC.

11. A nationalistic motivation can be found in the origins of this institute because Peruvian physiologists tried to demonstrate the capability of native Indian highlanders in overcoming the effects of an environment with little oxygen. See Marcos Cueto, "Andean Biology in Peru: Scientific Styles on the Periphery," *Isis* 80 (1989): 640–59.

12. McCann to Gregg, March 1, 1937, RFA, R.G. 1.1, Series 331, Box 1, Folder 6, RAC.

13. Regarding the scientific disruption caused by the Second World War, see the report prepared for the Division of Foreign Relations of the U.S. National Research Council, Walter B. Cannon and Richard M. Field, "International Relations in Science, A Review of Their Aims and Methods in the Past and in the Future," *Chronica Botanica* 9 (1945): 255–98.

14. "Harvard to Test Soldiers for Fatigue," *The Boston Daily Globe*, April 24, 1941, newspaper clipping, Folder Harvard Fatigue Laboratory, Baker Library, Manuscripts and Archives Department, Harvard Business School.

15. See Fernando Henrique Cardoso and Enzo Faletto, *Dependency and Development in Latin America* (Berkeley: University of California Press, 1979), 128–29.

16. According to a study, the Good Neighbor policy was trapped in a contradiction. On the one hand it showed respect, previously nonexistent for Latin American governments and culture; "on the other hand, it manifested itself as an effort to further subordinate them to the United States." Alonso Aguilar, *Pan-Americanism, from Monroe to the Present: A View from the Other Side* (New York: Monthly Review Press, 1968), 70.

17. For an illuminating account of the post–World War II relationship between the United States and Latin America, see Stephen G. Rabe, *Eisenhower and Latin America: The Foreign Policy of Anticommunism* (Chapel Hill: The University of North Carolina Press, 1988).

18. Hull to Cannon, September 28, 1939, Walter B. Cannon papers (hereafter Cannon papers), Box 66, Folder 880, Harvard Medical School, Boston, Francis Countway Library (hereafter FCL).

19. This new direction was reinforced by the fact that, since the early 1930s, the RF had operated under a program emphasizing the application of physics and chemistry to experimental biology. See Robert E. Kohler, "Warren Weaver and the Rockefeller Foundation Program in Molecular Biology: A Case Study in the Management of Science," in Nathan Reingold, ed., *The Sciences in the American Context: New Perspectives* (Washington D.C.: Smithsonian Institution Press, 1979), 243–93.

20. "Report of the Committee of Review, Appraisal and Advice. The Rockefeller Foundation, Division of Natural Sciences, 1939," Cannon papers, Box 92, Folder 1269, FCL.

21. Minutes of the first meeting of the Inter-Divisional Committee on Latin America, November 29, 1943, RFA, R.G. 1.2, Series 300, Box 1, Folder 7, RAC. No effort will be made here to discuss the activities of this committee, because it encompassed all countries and areas of action of the RF.

22. For a biographical account of the first years of Cannon, see Saul Benison, A. Clifford Barger, and Elin L. Wolfe, *Walter B. Cannon: The Life and Times of a Young Scientist* (Cambridge: Belknap Press of Harvard University Press, 1987). A historical analysis of U.S. physiology appears in Gerald Geison, ed., *Physiology in the American Context 1850–1940* (Bethesda, Md.: American Physiological Society, 1987).

23. Walter B. Cannon, *Curso de Fisiología de Laboratorio, Versión Española de la 6a Edición Inglesa* (New York: Appleton & Cia, 1929). The information on the free distribution appears in *Breve Información Bio-Bibliográfica del Coronel Médico Cirujano José Joaquín Izquierdo* (Mexico, n.p., 1945), 10. Other scientific books of Cannon's published in Latin America by his disciples were *La Sabiduría del Cuerpo* (Mexico: Editorial Séneca, 1941); *Digestión y Salud* (Buenos Aires: Emece Editores S.A., 1945); and *La Sabedoria do Corpo* (Sao Paulo: Companhia Editora Nacional, 1946).

24. Cannon to Lewis, May 31, 1933, Cannon papers, Box 65, Folder 864, FCL.

25. An officer of the RF wrote to Cannon, "such aid is envisaged if, as you point out, it is an essential item for the continuation of the techniques and methods which they learnt abroad"; O'Brien to Cannon, June 14, 1938, Cannon papers, Box 93, Folder 1282, FCL.

26. This move was confirmed by the Director of the Division of Medical Education; see Alan Gregg to Cannon, October 26, 1944, Cannon papers, Box 93, Folder 1288, FCL.

27. These reports are discussed in my "Visions of Science and Development: The Rockefeller Foundation's Latin American Surveys of the 1920s," chapter 1 of the present volume.

28. Abraham Flexner, *Medical Education in the United States and Canada, Bulletin No. 4* (New York: Carnegie Foundation for the Advancement of Teaching, 1910).

29. Medical Schools of Latin America, RFA, R.G. 1.1, Series 300, Box 1, Folder 4, RAC.

30. Lambert to Janney, March 15, 1943, RFA, R.G. 1.1, Series 309, Box 5, Folder 40, RAC.

31. See David Rock, *Argentina 1516–1987: From Spanish Colonization to Alfonsin* (Berkeley: University of California Press, 1987), 162–249.

32. James R. Scobie, *Buenos Aires: Plaza to Suburb, 1870–1910* (New York: Oxford University Press, 1974).

33. A Laboratory Grows in Argentina, RFA, R.G. 1.2, Series 301, Box 1, Folder 4, RAC.

34. *Memoria del Instituto de Biología y Medicina Experimental* (Buenos Aires: Talleres Gráficos de Sebastián Amorrortu e Hijos, 1946), 4.

35. In 1947 the Nobel Prize in Physiology or Medicine was divided between Houssay and Carl and Gerty Cori of the United States.

36. A complete review of the foreign researchers who worked at the Instituto de Biología y Medicina Experimental appears in a letter from Houssay to Oliver, September 30, 1954, RFA, R.G. 1.2, Series 301, Box 2, Folder 10, RAC.

37. For a historical account of the first years of this journal, see Hebe Vessuri, "Una Estrategia de Publicación Científica para la Fisiología Latinoamericana: Acta Physiologica Latinoamericana, 1950–1971," *Interciencia* 14 (1989): 9–13.

38. Watson to Warren, July 16, 1954, RFA, R.G. 1.2, Series 301, Box 2, Folder 10, RAC.

39. Warren to Watson, August 2, 1954, RFA, R.G. 1.2, Series 301, Box 2, Folder 10, RAC.

40. Harry M. Miller, Jr., "Observations on Argentina–1957," RFA, R.G. 1.2, Series 300, Box 2, Folder 8, RAC.

41. Houssay to Miller, March 3, 1959, RFA, R.G. 1.2, Series 301, Box 2, Folder 11, RAC.

42. Janey to Miller, January 19, 1959, RFA, R.G. 1.2, Series 301, Box 2, Folder 11, RAC.

43. For a brief history of the institute and its significance for Latin American medical

science, see Ignacio Chavez, *El Nuevo Instituto Nacional de Cardiología* (Mexico: Imprenta Madero, 1978).

44. One of the main products of their work was Walter B. Cannon and Arturo Rosenblueth, *Autonomic Neuro-effector Systems* (New York: Macmillan, 1937). For an excellent account of Rosenblueth and Cannon's relationship, see Louisa Barclay Benton, "Arturo Rosenblueth, Success or Failure: A Consideration of the Forces Which Lie behind 'Success' and 'Failure' in Science through the Biography of a Mexican Neurophysiologist" (Bachelor's thesis, Harvard University, 1986), Harvard University Archives.

45. Cannon to Gregg, September 27, 1944, RFA, R.G. 1.1. Series 323, Box 13, Folder 97, RAC.

46. See Michael C. Meyer and William L. Sherman, *The Course of Mexican History* (New York and Oxford: Oxford University Press, 1983), 596–651.

47. Visit to Mexico, March 1–14, 1941, RFA, R.G. 1.1, Series 323, Box 13, Folder 95, RAC.

48. See the chapter, "Society and Culture since World War II," in Meyer and Sherman, *Mexican History*, 688–704.

49. National Institute of Cardiology, Mexico City, RFA, R.G. 1.2, Series 301, Box 2, Folder 10, RAC.

50. Robert A. Lambert, Diary, December 18, 1945, RFA, R.G. 1.1, Series 323, Box 13, Folder 97, RAC.

51. Interview, February 7, 1946, RFA, R.G. 1.1, Series 323, Box 13, Folder 97, RAC.

52. National Institute of Cardiology, Mexico City, RFA, R.G. 1.1, Series 323, Box 13, Folder 97, RAC.

53. Ibid.

54. Ibid.

55. See Norbert Wiener, *Cybernetics, or Control and Communication in the Animal and the Machine*, Second edition (Cambridge, Mass.: M.I.T. Press, 1961).

56. Rosenblueth to Cannon, May 29, 1944, Cannon papers, Box 117, Folder 1626, FCL.

57. Chavez to Warren, September 22, 1953, RFA, R.G. 1.1, Series 323, Box 14, Folder 100, RAC.

58. Benton, "Rosenblueth: Success or Failure," 110.

59. Novis to Miller, June 20, 1955, RFA, R.G. 1.1, Series 305, Box 11, Folder 97, RAC.

60. Simoes to Miller, December 24, 1958, RFA, R.G. 1.2, Series 305, Box 9, Folder 83, RAC.

61. Watson to Miller, October 28, 1955, RFA, R.G. 1.2, Series 305, Box 9, Folder 83, RAC.

62. J. L. Hydrick, "Report on the Programs of the International Health Division of the Rockefeller Foundation in Ecuador and Peru, January–December 1948," in International Health Division of the Rockefeller Foundation, The Rio de la Plata and Andean Region, Annual Report, 1948, 97–98, Lewis Wendell Hackett Collection, Acc. No. 33, Box 19, RAC.

63. This is confirmed in a letter from Watson to Bugher, May 26, 1955, RFA, R.G. 1.2, Series 305, Box 4, Folder 33, RAC.

64. There were other Brazilian physiologists conducting interesting research during the 1950s and early 1960s. Among them were Luiz Carlos Junqueira from Sao Paulo and Carlos Chagas from Rio de Janeiro. For an account of Brazilian physiology research during the twentieth century, see Jose Ribeiro do Valle, "Alguns Aspectos da Evolução da Fisiologia no Brasil," in Mário Guimarães Ferri and Shozo Motoyama, eds., *História das Ciências no Brasil* (São Paulo: Editora da Universidade de São Paulo, 1979) 1: 151–74.

65. Report by Robert Briggs Watson, November 26, 1958, RFA, R.G. 1.2, Series 305, Box 4, Folder 33, RAC.

66. RFA, R.G. 1.2, Series 300A, Box 2A, Folder 16D, RAC.

67. For an account of this period, see Thomas E. Skidmore, *Politics in Brazil 1930–1964: An Experiment in Democracy* (New York: Oxford University Press, 1986).

68. Watson to Weaver, March 20, 1962, RFA, R.G. 1.1, Series 305, Box 4, Folder 33, RAC.

69. See Cueto, "Andean Biology."

70. Trustees' Confidential Report, October 1, 1954, RFA, R.G. 1.1, Series 331, Box 1, Folder 8, RAC.

71. Institute of Andean Biology, April 20, 1951, RFA, R.G. 1.1, Series 331, Box 2, Folder 15, RAC.

72. See Cueto, "Andean Biology."

73. For an account of economic policies and political developments in Peru in the twentieth century, see Rosemary Thorp and Geoffrey Bertram, *Peru 1890–1977: Growth and Policy in an Open Economy* (New York: Columbia University Press, 1978).

74. Ibid.

75. Hurtado to Castro de Mendoza, September 6, 1954, Folder C, Memorial Hurtado, Instituto de Investigaciones de Altura, Universidad Peruana Cayetano Heredia, Lima, Peru.

76. Herman Rahn, University of Rochester Medical School, Travel Grant, October 5, 1955, RFA, R.G. 1.1, Series 200A, Box 159, Folder 1441, RAC.

77. Houssay to Cannon, Cannon papers, January 2, 1923, Box 65, Folder 857, FCL. A few years later, Houssay learned to speak and write in English.

78. Houssay to Cannon, November 22, 1943, Cannon papers, Box 65, Folder 860, FCL.

79. Houssay to Dr. Chandes, February 10, 1949, Memorial Hurtado, Instituto de Investigaciones de Altura, Universidad Peruana Cayetano Heredia. Houssay's ideas regarding science and nationalism appear in Bernardo Houssay, *Obra Realizada, Principios que la Guiaron, Ensenanza-Investigación-Medicina* (Buenos Aires: Consejo Nacional de Investigaciones Científicas y Técnicas, 1958).

80. Lewis to Cannon, December 4, 1940, Cannon papers, Box 65, Folder 864, FCL.

81. Robert A. Lambert sent a copy of the letter to Cannon; Hurtado to Lambert, January 4, 1937, Cannon papers, Box 93, Folder 1280, FCL.

82. According to Cannon, the influence of U.S. researchers in Latin America was going to provide "the same sort of stimulating service which European leaders in the medical sciences performed for ambitious American doctors two or three generations ago"; Walter B. Cannon, "Problems Confronting Medical Investigators," *Science* 94 (1941): 177.

83. Lewis to Cannon, January 22, 1932, Cannon papers, Box 65, Folder 863, FCL.

84. Benson, "Rosenblueth: Success or Failure," 64.

85. Lambert to Taylor, August 25, 1943, RFA, R.G. 1.1, Series 301, Box 3, Folder 32, RAC.

86. University of Rio Grande do Sul, Brazil, Faculty of Medicine, RFA, R.G. 1.2, Series 305, Box 15, Folder 133, RAC.

87. Watson to Weir, May 9, 1955, RFA, R.G. 1.2, Series 305, Box 7, Folder 58, RAC.

88. For an analysis of the role played by the RF in one of these schools, see the history of Colombia's Universidad del Valle, Guillermo Orozco, *De la Escuela de Medicina y su Universidad* (Cali: Universidad del Valle, 1984).

89. For an analysis of this problem, see José Joaquín Brunner, *Universidad y Sociedad en América Latina: Un Esquema de Interpretación* (Caracas: CRESALC, 1985).

90. See Saul Franco Agudelo, "The Rockefeller Foundation's Antimalarial Program in Latin America: Donating or Dominating?" *International Journal of Health Services* 1 (1983): 51–57; María Eliana Labra, "O Movimento Sanitárista Nos Anos 20, Da Conexão Sanitária Internacional a especialização em Sáude Pública no Brasil" (Master's thesis, Escola Brasileira de Administraçáo Pública, Rio de Janeiro, 1985).

91. See, for example, Kenneth M. Ludmerer, *Learning to Heal: The Development of American Medical Education* (New York, Basic Books, 1985).

7

The Rockefeller Foundation and the Emergence of Genetics in Brazil, 1943–1960

Thomas F. Glick

ALTHOUGH CLASSICAL GENETICS slowly diffused through Brazilian agricultural and medical schools in the 1920s and 1930s, its definitive implantation began only in 1943 when Theodosius Dobzhansky began research in Brazil. Dobzhansky, a Russian biologist who emigrated to the United States in the 1920s, introduced a circle of Brazilian disciples to the methods of *Drosophila* genetics and trained its members in the rudiments of population biology and neo-Darwinian evolutionary theory. This introduction, broadened during Dobzhansky's sabbatical in Brazil in 1948–49, provided the fledgling Brazilian geneticists with a solid evolutionary research program and ready-made outlets, through Dobzhansky, for the publication of its results, in which Dobzhanksy participated. This entire process took place under the aegis of the Rockefeller Foundation, which provided funding for equipment, for travel grants, and for research fellowships for the Brazilian biologists. In total, it represents a model success story in Latin American science and illustrates the creative intervention of a U.S. foundation in the development of a scientific discipline abroad.

Brazilian Genetics B.D. (Before Dobzhansky)

All the early teachers of genetics in Brazil were self-taught in that field. Carlos Teixeira Mendes, professor of agriculture at Piracicaba, was the first Brazilian to teach Mendelian genetics. His 1917 thesis on the improvement of agricultural varieties was based in part on De Vries, and he began to teach genetics the following year. He wrote extensively on selection, one of his common themes being that empirical selection for plant and animal breeding, which had been typically understood in Lamarckian terms, could in fact be explained by Darwin.[1] Two colleagues of Teixeira Mendes's also taught genetics in Piracicaba. One, Octavio Domingues, was professor of "zootechnics"; by 1930 he was up-to-date on the *Drosophila* studies of Morgan's school.[2] The other, zoology professor Salvador de Toledo Piza, Jr., was opposed to the concept of a material gene. Genes in his view were neither particulate nor capable of self-replication, but simply the expressions of "plastinemes" or filaments of the cell nucleus.[3] Piza did play a positive role in the diffusion of genetics, however, by standardizing its technical terminology in Portuguese.[4]

Another autodidact, one who was to play a signal role in the institutionalization of genetics in Brazil, was André Dreyfus, who began teaching genetics at the medical school in Sao Paulo in the late 1920s. He may have been the first Brazilian to have a grasp of the neo-Darwinian synthesis since he was citing R. A. Fisher by 1929, the same year in which at a eugenics congress in Rio de Janeiro he told the eugenicists that Lamarckism must be abandoned, including their favored notion that a good environment would favorably affect a person's hereditary makeup.[5]

Mainstream genetics arrived in force in the mid-1930s, when the German refugee Friedrich Brieger was named professor of cytogenetics at Piracicaba and when Carlos Krug, trained at Cornell, became director of the genetics service (founded in 1929 with Edgard Taschdian as its first director) at the Instituto Agronômico at Campinas. Krug had taught genetics since 1933, and Brieger began teaching in 1936. Brieger in the following decade produced an important series of articles on the evolution of orchids, and Krug's research was on the genetics of coffee. In the early 1940s, both Krug and Brieger published important accounts of the neo-Darwinian synthesis and the role of genetics in the evolutionary process.[6]

In 1934 Dreyfus became the first professor of biology at the newly founded University of Sao Paulo, where from the beginning of his tenure he gave courses on genetics and evolution. The university itself was an experiment in creating an "American-style" university in Latin America. Because it adopted from the start a "full-time" administrative regime for professors and researchers, it enjoyed broad support from the Rockefeller Foundation, one of whose fixations with regard to science in Latin America was an absolute insistence on full-time positions so that scientists would not have to have multiple jobs in order to earn a living. When in 1941 Dreyfus attracted the attention of one of the Rockefeller's field men in Latin America, the stage was set for the implantation of modern genetics.

The Apostolate

The chain of events that brought Theodosius Dobzhansky to Brazil began when a foundation official, Harry Miller, "made his first trip below the Caribbean in 1941 on a prospecting tour of the South American universities, looking for opportunities to assist teaching and research in the natural sciences."[7] On the trip, he met Dreyfus and offered him a fellowship to study in the United States. Dreyfus was concerned lest no one teach his classes, and Miller promised to find an American to fill in. Miller thought that Dobzhansky would fill the bill admirably. On November 18, 1942, Dobzhansky called Miller to discuss "his desire to collect Drosophila species in the Amazon Valley, one of the unusual regions in the world where no seasonal climatic changes take place." Several days later, he further specified that he would like "nothing better than to spend two or more months in the Amazon Valley" and go as visiting professor to Dreyfus's department. The following day, Miller noted that he had written Dreyfus to sound him out on the developing plans, noting that "a further point in favor of this exchange is that the subject of genetics is not too well represented in Brazil," the only names coming to mind being those of Brieger and Krug.[8]

On the 24th, Dobzhansky had already prepared and sent to Miller a rationale for the Brazilian excursion, in the form of a memorandum entitled "Population Genetics in Tropical Latin America." After reviewing advances in quantitative and theoretical genetics by Wright, Fisher, Haldane, and Muller, Dobzhansky observed that all studies of population genetics thus far had been confined to temperate-zone species of the northern hemisphere. All temperate zones have sharp seasonal changes, and thus organisms must be adapted to several different environments in their lifetimes. But nothing was known of population structures in tropical counties, where there is scant climatic variation. It was important, therefore, to find out

> whether or not in typically tropical species the mutation rates are low,
> the migration rates reduced to a minimum, [whether] the
> diversification of the populations in local races is high, and proceeding
> to a considerable extent along non-adaptive lines.

Only actual experiments can find this out.[9] He would sample collections of Brazilian *Drosophila*, choose two or three for intensive study, and ship cultures to New York for further work. Dobzhansky was later to admit that his initial hypothesis, which presumed that without seasonal fluctuations there would be no genetic drift, was wrong.[10]

What emerged in early 1943 as Dobzhansky's "Grand Plan" was a continuation of his famous series of articles on the genetics of natural populations, under tropical conditions, in Brazil, to compare those results to his studies of temperate species, mainly of *Drosophila*. A bit of background will place this plan in its historical context. Dobzhansky's overriding interest in *Drosophila* genetics was the study of microevolution. As early as 1922, he had realized that to understand it he would have to acquire information about "geographical distribution and variability between populations, individual variability within populations and the inheritance of variability."[11] This could not be done under classical techniques in a laboratory (like Morgan's, where Dobzhansky worked for many years), because the *Drosophila* species studied were intersterile. In order to get the information he required, natural populations would have to be studied. Studies of *D. pseudoobscura* by various geneticists in the 1920s clarified the way to do that. This species and a sibling species, *D. persimilis*, were morphologically identical but reproductively isolated, had many genes in common, and also exhibited considerable chromosomal variation throughout the population. Studies of natural populations of *pseudoobscura* offered the possibility of constructing phylogenies of the strains studied and changed the locus of microevolutionary studies from the origin of species to that of races.[12] The evolutionary significance of the study of this kind of population became clearer to Dobzhansky the more he came to understand Sewall Wright's theories of selection: The most flexible and efficient means of evolution in nature, according to Wright's view, came when a large species was subdivided into many small local races with some migration, and the effects of both selection and random drift were intermediate. Selection would then operate at both the intra- and intergroup levels.[13] As soon as Dobzhansky realized that microevolution could best be studied by comparing races of *Drosophila*, he threw himself into collecting samples of natural populations, cutting a wide geographical path in the American west from Guatemala to Alaska.

It was the opportunity to make such collections of Brazilian species of *Drosophila* that was, initially, the main attraction for Dobzhansky. Still believing that the main interest in studying tropical *Drosophila* was to test for genetic drift in a region of scant seasonal variation, Dobzhansky wrote to Sewall Wright in January 1943, inviting him to New York for a discussion of Dobzhansky's hypothesis. Wright could not make the trip, and some months later Dobzhanksy traveled to Chicago by train for a two-day meeting. The meeting must have been fruitful, for Dobzhansky soon dropped the focus on genetic drift, but the details of their conversation are not recorded.[14]

Once in Brazil in 1943, Dobzhansky toured the countryside tirelessly with a young geneticist, Crodowaldo Pavan, experiencing a true "Columbian" discovery of America, which seemed to him to be a worldly paradise (if *Drosophila* may be likened to gold!). Consider this description of *Drosophila* swarms on an island in the Rio Negro (in 1948, but no doubt an accurate representation of his early Amazonian field trips as well):

> The jungle swarms with *Drosophila*. We landed on an island in the
> Rio Negro and found the ground covered with fallen fruits of many
> species of trees, and on these fruits there were clouds of *Drosophilae*

the equal of which I have not seen in my fifteen years of
Drosophila-collecting in nature, either on natural or on artificial bait.
There were surely hundreds of individuals per square meter on this
island, which in my notes is called the Drosophila Paradise Island. But
even outside of this island phenomenal population densities have
repeatedly been found. And there exist in all collections quite a lot of
different species, several times the number one finds in the Californian
Sierra. The fly (not *Drosophila*) in the ointment is that at least half of
these species are new, and we cannot find time to describe them. Yet
how does one handle these observations if even some of the
commonest species one finds have no names?[15]

In 1943 Dobzhansky discovered twenty-three new species of *Drosophila* in a
brief Amazonian expedition, an accomplishment which (in the words of a later
Brazilian report) "set Brazilians to collecting for themselves. Moreover, it sent
them back to their home institutions at the end of the summer with an almost
religious zeal for the new science".[16]

Dobzhansky's course was just as successful or more. As one participant
recalled it:

Dobzhansky came for three months in 1943 and gave an excellent
course on evolution in Portuguese. Dreyfus, Pavan and Brito (then a
second-year student) translated the lectures which were then
published by the Ministry of Agriculture. Dobzhansky had studied
Portuguese for three months and was able to speak it quite well, since
Russian sounds are similar.
    About twenty students attended, but most of the University of Sao
Paulo biologists also went to the course as well as people from the
Instituto de Biologia, Veterinary School—Rocha Lima, Clemente
Pereira, Zeferino Vaz etc. So the course influenced all the biologists of
Sao Paulo. Krug, Brieger, and Alcides Carvalho came from Piracicaba
and some from Campinas came as well.[17]

During the three months Dobzhansky stayed in Brazil in 1943, as he later
recalled, Dreyfus

took me under his wing, and decided to use my visit as a means of
building up his department of biology, starting genetic research. Being
a man of great energy, he did what he intended to do. A school of
genetics appeared in Brazil, a school which did not exist before, and it
is a real disappointment to me that this school did not develop, later,
after Dreyfus's death, as well as it might have developed.[18]

Dobzhansky's overly pessimistic assessment twenty years later should not obscure Dreyfus's use of him to generate interest in genetics in Brazil, and Dobzhansky's full assent to the project.

The Biologists Organize

Following A. G. Cavalcanti, one of Dobzhansky's trainees of the F_1 generation, *Drosophila* research in Brazil can be comprehended in three phases.[19] The first phase (1943–48) was wholly devoted to systematics and morphology. In terms of research, the period begins with Dobzhansky and Pavan's 1943 study on Brazilian species of *Drosophila* and ends with Pavan and Antonio Brito da Cunha's 1947 article on the same topic.[20] In Brazil Dobzhansky identified two species that seemed interesting to study, the idea being to compare a species with a large population with another of small population. Those selected were *D. willistoni* and *D. prosaltans*, samples of which he took back to his laboratory in New York where he and Boris Spassky studied them and prepared experimental strains.[21] In Brazil, *Drosophila* work continued at Sao Paulo, and in 1944 Cavalcanti began a similar group in Rio de Janeiro with Chana Malogolowkin and Oswaldo Frota-Pessoa. During the same period, a series of Brazilians passed through Dobzhansky's laboratory at Columbia: Pavan first, then Brito da Cunha, Cavalcanti, and Warwick Kerr (and later, Malogolowkin, Antonio Cordeiro, and Frota-Pessoa). "Usually we had at least one Brazilian in the laboratory, seldom two and seldom none," Dobzhansky later recalled.[22]

Next came the experimental phase, which began with Dobzhansky's second visit in 1948–49. This stage was almost wholly devoted to studying the genetics of natural populations of the two *Drosophila* species, *willistoni* and *prosaltans*, a large group and a small one, respectively. This phase involved the field training of the geneticists, some of whom had already worked with him in New York, and who soon after created the institutional structure of Brazilian genetics.

When Dobzhansky returned for his sabbatical year in 1948–49, he established two working groups in Sao Paulo. The first group, composed of Dobzhansky, Pavan, Malogolowkin, and Cavalcanti, studied the genetic variability of *D. willistoni* and *D. prosaltans*. This study showed that the frequency of lethal genes in natural populations is higher in tropical than in temperate regions. The second group, which included Dobzhansky, Brito da Cunha, Cordeiro, Hans Burla from Zurich, and Marta Wedel from Buenos Aires, studied chromosomal variability.[23] This group studied four sibling species of *Drosophila* (*willistoni*, *paulistorum*, *tropicalis*, and *equinoxialis*). Different patterns were found in the chromosomes of the salivary glands. When the researchers tried to cross the species, they found them to be intersterile, even though they were morphologically very similar. The pattern of the frequency of inversions in these species showed Darwinian selection. The degree of variability of the species correlated with diversity of environment. *Willistoni*, in a dry area of Bahia, had a low frequency of inversion, but in a more diverse environment, a high frequency. It was also found that *willistoni* was ecologically more versatile than *tropicalis*, which had much less chromosomal variability.

The comparison also revealed a significant point about the relationships between competition and the origin of species. In the Amazon basin, a region that presents a number of very similar niches for *Drosophila*, the closest competitors are sibling species. In those places where *willistoni* is abundant, variability is great. High variability conveys the capacity to exploit more niches, especially in view of competition with sibling species.[24] By the end of the year, more that forty different populations had been studied, an achievement that took Dobzhansky and his Brazilian colleagues to many parts of the country, an odyssey described in Dobzhansky's fascinating travel reports, published posthumously as *The Roving Naturalist*. Travel was by boat and air, the planes, plus all-important official introductions, being supplied by the Brazilian air force, thanks to Cavalcanti's father, a general.

When Miller asked Burla to evaluate the year's research, the Swiss replied,

> It is much more hard for me to get any clear idea about the effect of your grant for genetics in Brazil itself. Both Pavan and Brito da Cunha are good and reliable research workers. Cavalcanti and Cordeiro are building up their laboratories, every success of their activity lies in the future, as in my case. The struggle against Brazilian burocracy [sic] and the necessity to take care of personal relations are much of a hindrance for the development of effective work, as you know certainly much better than I do. All the same I think that most conditions to work are now favorable in Brazil. The workers have their program, they saw an example how to behave, they possess excellent optical equipment for research work, laboratories with temperature rooms (in Sao Paulo) and assistant salaries which are higher than in most European countries.[25]

Significance of the 1948–49 Research

In an admirably concise report published in 1949, Dobzhansky summarized this phase of the research.[26] The keynotes of this research were variation and adaptation ("the principal driving force of organic evolution"). The study of tropical *Drosophilae* was designed to explore the differing tempo and "creativeness" of evolution in the tropics, as compared to the same processes in temperate climates. Beginning with his 1943 survey with Pavan, Dobzhansky and his group had collected and examined 110,000 individuals sampled from thirty-five ecologically diverse sites in Brazil.[27] The extreme diversity of *Drosophila* species in small neighborhoods (as many as thirty species in areas of less than one square kilometer) suggested that tropical environments contained "a greater variety of ecological niches (biological opportunities)" than did temperate environments.[28] Adaptation to this diversity of niches could occur in two ways: either numerous species will adjust to the same niche or a small number of species will become polymorphic in order to exploit a variety of niches.

Drosophila, a particularly labile species, had followed both options, and experiments were devised to illustrate each process. First Burla, Brito da Cunha, Cavalcanti, and Pavan had demonstrated microterritorial differentiation, whereby a community consisting of many competing species was broken up "into tiny colonies with a much smaller number of competitors."[29] This study and others followed up on the observation that different *Drosophila* populations displayed different food choices. Pavan, Brito da Cunha, and Dobzhansky, collecting *Drosophila* on its natural foods (fermenting fruits) showed that the flies displayed preference for different foods but without any rigid dietary specialization. Dobzhansky was not clear on how to account for the variations in food attractiveness but concluded that this flexibility alleviated competition.[30] In the study cited above, Burla et al. had performed diffusion-rate experiments by releasing flies with a mutant trait at a site near Sao Paulo, to compare the number of mutants and wild flies at given locations. The experiment showed a propensity of *Drosophila willistoni* to form semi-isolated colonies. These experiments, designed to illustrate the population dynamics of *Drosophila*, led to the conclusion that aggregate population densities were greater in the tropics than in the temperate forests of California, where Dobzhansky had collected comparable data.[31] Similar species shared niches, taking advantage of the great variety of food available in tropical forests.

Another set of experiments was designed to illustrate concealed genetic variability in the form of chromosomal polymorphism and to explain the adaptive advantages that such variability conveyed. Chromosomal polymorphism refers to the phenomenon of inversions of chromosome segments. Here the key study was one by Brito da Cunha, Burla, and Dobzhansky on chromosomal polymorphism in *willistoni*,[32] which sought to explain the significance of the very high incidence of chromosomal inversions in this species, thirty-four different inversions—"the most inverted species known," as Dobzhansky wrote to Sewall Wright (*pseudoobscura*, the most-studied temperate species, had only twenty such inversions).[33] Dobzhansky sought to relate this remarkable genetic variability to the observed flexibility of *willistoni* in selecting fruit to eat. The ecological versatility of the species was correlated with genetic variability: "The amount of adaptive polymorphism present in the population of a species in a given region is, in general, proportional to the variety of habitats which the species has mastered." Thus, chromosomal polymorphism was a means "whereby the species becomes capable of occupying and exploiting efficiently a variety of ecological niches in the environment in which it lives."[34] Other researchers soon tested this hypothesis, with discrepant results, and so Brito da Cunha and Dobzhansky published a further study to clarify it. Polymorphism, they reiterated, was "a method which a population can use to become adapted to a multiform environment".[35] The inversions were maintained in the population ("balanced polymorphism") owing to the adaptive advantage conveyed to heterozygotes. Chromosomal polymorphism explained the ability of *willistoni* and its three sibling species to exploit very similar niches.

From Dobzhansky's perspective in 1950, chromosomal polymorphism illustrated the key feature of evolution in the tropics, namely "adaptive versatility." One fact that had brought him to this conclusion, interestingly enough, was his recognition that "the widespread opinion that seasonal changes are absent in the tropics is a misapprehension."[36] The fact that tropical environments were changeable meant the variety of ecological niches and complexity of ecological processes was even richer than at first he supposed. Tropical environments thus provide more "evolutionary challenges" than do their temperate counterparts. The responses of organisms "constitute progressive evolution".

Several years later, in an address at the Brazilian Academy of Sciences in Rio de Janeiro, Dobzhansky reflected on the nature of adaptation in the light of the *willistoni* research. The morphological traits used to classify *Drosophila* appeared to be adaptively neutral. The reproductive isolation of *willistoni* and its siblings was not attributable to any single trait but to "syndromes" of genes. Genotypes, not single genes, are subject to natural selection, and their adaptive value is made manifest in the fitness of the phenotype. The adaptiveness of a trait or group of traits, therefore, "cannot be considered apart from the developmental pattern that facilitates its survival in a certain succession of environments."[37]

This set of experiments, neatly framed in print by the two articles by Dobzhansky just cited, had considerable staying power in the literature. Three of the articles are cited by Sewall Wright in his "summa" on evolution and population genetics with reference to frequency distributions and variability within and among natural populations.[38] We may also note the propensity for citations of certain of these articles (as recorded in *Science Citation Index*) to increase over time. Thus Brito da Cunha, Burla, and Dobzhansky's 1950 article on chromosomal polymorphism[39] was cited ten times in 1965–69, eight in 1970–74, seven in 1975–79, only to rise to fourteen in 1980–84; it was still cited in 1990 (thrice). Similarly, Brito da Cunha and Dobzhansky's 1954 follow-up article[40] was cited seven times in 1965–69, ten in 1970–74, seven in 1975–79, eighteen in 1980–84 (and, again, thrice in 1990).

Armageddon: Angra dos Reis

Dobzhansky made a third trip in summer 1952 to Belem and a short stop in 1953 to attend as an examiner Pavan's examination for a professorship. The fifth and last trip was another sabbatical in 1955–56, "the least pleasant trip of the whole bunch," in Dobzhansky's view, "an ill-starred expedition."[41] Dobzhansky, viewing by air some tropical islands in the bay of Angra dos Reis, off Rio, had an ingenious, even grandiose idea: "A temptation arose almost as soon as I saw it for the first time to utilize some of these green islands as giant population cages for experiments on *Drosophila*."[42]

The first experiment Dobzhansky devised was to see whether chromosome inversions were adaptive, by releasing into the wild continental flies with inversions that did not exist among the island population, to see if the inversion would be maintained in the environment, for how long, in what frequency, and so forth. "Just exactly what happened," according to Dobzhansky,

> we will never find out. I strongly suspect that one of the graduate students who was given a sort of subsidiary problem, which was to release flies of a different species . . . (I hope by mistake) released *D. willistoni*. In short the whole thing caved in.[43]

Indeed, the inversions in the released population already existed in the natural population. What Dobzhansky thought was a new inversion was not new. The Brazilians had a different explanation of what went wrong. Pavan and Brito da Cunha thought the initial assumption had been wrong because the continental population had not been adequately studied. Dobzhansky thought some contamination of the sample population had occurred and decided that a student had mixed up the populations. The Brazilian professors defended the student, and that was the end of Dobzhansky's involvement with Brazilian genetics.[44]

The project, however, continued and produced significant results, some of which were published under Dobzhansky's name, so his pique must not have been as strong as his memoirs later indicated. The results of a second experiment were in fact interesting: no differences between x-ray-induced and naturally occurring lethal genes were detected in heterozygous conditions. "This result was unexpected because for a long time it was believed that wild lethals, being subject to natural selection for a long time, would be less deleterious in heterozygous conditions, than recently induced lethals."[45] These results seemed to indicate that heterozygism per se has an adaptive value.

The Third Phase

By 1956 there were nearly a dozen centers of genetics research in Brazil, several of which, including Cavalcanti's in Rio de Janeiro and Cordeiro's in Porto Alegre, founded in 1949, had begun as *Drosophila* research centers. Cavalcanti's third phase, that of independently conceived research, was already underway, representative projects being Cavalcanti's on sex ratio in *D. prosaltans*, Malogolowkin's on genital characters in the systematics of *Drosophila*; Warwick Kerr's (in conjunction with Sewall Wright) on the effect of genetic drift on the structure of a *Drosophila* population, *melanogaster* in this case; Francisco Salzano's (in conjunction with Hampton Carson) on factors causing sexual isolation in natural populations of *Drosophila*, not to mention a host of projects having nothing to do with *Drosophila*.

When considering the success of Brazilian genetics, a number of factors must be considered. First, as Dobzhansky noted in eulogizing André Dreyfus, "A scientific school, if it is based on an important and viable idea, can live on for a long time after the death of its founder."[46] The research program itself is the main reason for a discipline's success. Moreover, the nature of the natural populations of *Drosophila* available in Brazil gave local researchers an advantage. *Melanogaster*, Dobzhansky reminded, can be studied anywhere in the world.

> But a biologist in Brazil has the enormous advantage of using as material for his work the Brazilian species of *Drosophila*, or animals of the tropical sea of plants of the closed fields of the tropical forests. . . . The flies of the genus *Drosophila* have shown themselves to be the most favorable material for studies of problems of population genetics and of evolution.[47]

The research program, which was Dobzhansky's, together with the wonderful local stocks of *Drosophila*, combined to produce what Robert Merton calls a "strategic research site," characterized by a program that is constantly unfolding, producing strings of novel results, and in this case too, with instant access to U.S. laboratories and mainstream publications.

The second noteworthy aspect of the Brazilians' success was the cooperative, coordinated style of research adopted. Very early on, according to Pavan, a deliberate decision was made to limit the number of projects.[48] The cooperative organization of research, moreover, made it possible continually to formulate generalizations that led to new, unforeseen research problems. In the same 1952 speech, Dobzhansky noted that anyone could publish a scientific article. What is more difficult is "to develop a coherent research program, directed towards an important objective, whose results not merely fill journal pages, but resolve or reveal phenomena as yet unknown."[49] That coherence was achieved by coordination and cooperation among a rather large core group of researchers. Pavan noted its advantages in 1954:

> We believe that this type of work yields considerably better results
> than could be obtained from a series of independent and
> uncoordinated efforts of several investigators working in isolation.
> Apart from a mass of data obtained, the work resulted in the
> formulation of some broad, although for the time being tentative
> generalizations of general biological interest. As so often happens in
> scientific research the solution of some problems has raised a number
> of new problems. We are anxious to exploit the possibilities opened by
> the results obtained so far, and to develop the work further.[50]

The Rockefeller Foundation, in the person of Harry Miller, liked this style and encouraged it because it tended to bud off new centers of research led by persons trained in the core program. He had this in mind even before Dobzhansky's second visit, whose purpose (Miller jotted down in a note to himself) was twofold: (1) "capitalize on Dobzhansky's presence (he is in favor)"; and (2) "stimulate *Drosophila* work elsewhere, in a virgin continent."

As a result, *Drosophila* research became the royal road to success in Brazilian genetics. Even those young researchers who hated *Drosophilae* worked on them anyway. Newton Freire-Maia's departure from the official line on *Drosophila* research underlay his problems with Dobzhansky in 1948: "Inasmuch as he studied species which did not present chromosomal inversions, a cold war ensued with the *big boss* [Dobzhansky] who, in general, criticized the study of 'domesticated or semidomesticated' species."[51]

Of course, the Rockefeller Foundation's backing of the whole enterprise was a not insignificant factor in its success: first sending Dobzhansky, then making it possible for any member of the 1948 group who wanted to spend a year in the laboratory of Dobzhansky or some other geneticist (the human geneticists gravitated toward James Neel at Michigan), and finally Miller's great flexibility and willingness to listen to what the Brazilians were saying. Brazilian geneticists later viewed Miller's distinctive modus operandi as having made a major contribution to the discipline's success. Initial requests made of him were always oral: he would visit a laboratory, then have a general discussion with staff as to what their material needs were. Several weeks later, he would make the grant, and only *afterward* did the grantee have to present a specific list of requirements, generally apparatus and reagents.[52] Miller's role in the allocation of fellowships for study abroad was also significant. As later described by Brito da Cunha,

This judicious policy has as its best example that of the Rockefeller Foundation and especially of its director Dr. Harry M. Miller Jr. to whom is owing its success, thanks to the wisdom with which he acted. Miller visited our laboratories once or twice a year. He spoke with all the researchers of the laboratories to see how research was going. He consulted with the personnel of the laboratories and interviewed each candidate. He only would make a grant to researchers already approved for the doctorate and who had a proven guarantee of employment on returning to the country. Afterwards he would maintain close contact with the grantee abroad, as well as with his professor. On returning to Brazil, the grantee received funding for acquisition of equipment which he needed to proceed with his work. In this manner he took abroad only very well selected personnel and who already had acquired what was possible here in Brazil.[53]

The Rockefeller Foundation, moreover, had a specific set of values and a general, if undefined, modus operandi, which was "to make the peaks higher" by identifying scientific talent already imbued with the scientific value system or representing a good chance for being inculcated with it. Dreyfus so noted in 1948 as he cast about for prospective members of the study team: the RF, he noted "nao está interessada em diletantes," no matter how brilliant. Only those who gave promise of continuing the work started under Dobzhansky would be considered.[54] Dobzhansky, of course, promoted and proselytized for the same values.

Finally, I assume that there was a relationship between the great vogue of eugenics in Brazil of this period and the climate affecting the reception of Mendelian genetics. Such a link might partially explain the predisposition of the future "Drosophilists" to genetics as an interesting and socially relevant field,[55] and would also cast light on the climate of opinion prevailing in the academic and governmental bureaucracies that would support or even demand the creation of specialized research units in the field.

Notes

The research for this chapter was made possible by National Science Foundation grant #DIR-8911149. An earlier version of the chapter, entitled "Establishing Scientific Disciplines in Latin America: Genetics in Brazil, 1943–1960," was delivered as a paper at the meeting on "Science and Discovery," Madrid, June 1991.
1. Carlos T. Mendes, "A Selecçao Empirica," *Revista de Agricultura de Piracicaba* 3, 11–12 (1928): 1–20; Carlos T. Mendes, "As Teorias de Evolução e a Agricultura," *Revista de Agricultura*

(Piracicaba) 6 (1931): 263–73. On Teixeira Mendes, see S. de Toledo Piza, Jr., "Carlos Teixeira Mendes, o Primero Professor de Genética no Brasil," *Revista de Agricultura* (Piracicaba) 38 (1963): 47–52.

2. See Domingues's articles, "As Teorias de Hereditariedade," *Revista de Agricultura de Piracicaba* 4 (1929): 95–103, a popularization of Mendel's theory; and "Teoria de Factores Mendelianos," *Revista de Agricultura* (Piracicaba) 5 (1930): 3–18.

3. Piza's theory was embodied in his book, *Localização dos Factores na Linina Nuclear como Base de uma Nova Teoria sobre a Hereditariedade* (Piracicaba: Escola Agricola Luiz de Queiroz, 1930).

4. Because of his efforts, Brazilian geneticists use the spellings *cromossoma* instead of *cromozoma*, *gametócito* rather than *gametocito*, and so forth; S. de Toledo Piza, "Cadeira de Zoologia de Escola Superior de Agricultura 'Luiz de Queiroz' de Universidade de São Paulo," *Atos do Primeiro Simpósio Sul-Americano de Genética* (São Paulo: Universidade de São Paulo, 1961), 191.

5. André Dreyfus, "O Estado Actual do Problema de Hereditariedade," *Primeiro Congresso Brasileiro de Eugenia: Atos e Trabalhos* (Rio de Janeiro, 1929), 87–97.

6. C. A. Krug, "Genetica e Evoluçao," *Revista de Agricultura de Piracicaba* 15 (1940): 271–88; F. G. Brieger, "Consideraçoes Sobre o Mecanismo da Evolução," *Anais da Escola Superior de Agricultura "Luiz de Queiroz"* 1 (1944): 177–203. Brieger had of course read Dobzhansky's influential book, *Genetics and the Origin of Species* (New York: Columbia University Press, 1941); he cited it, along with R. A. Fisher, *The Genetical Theory of Natural Selection* (Oxford: Oxford University Press, 1930) in his bibliography.

7. "Grants from the Foundation: A Fundamental Study of Evolution," June 1, 1956, Rockefeller Foundation Archives (hereafter RFA), R.G. 1.2, Series 305, Box 45, Folder 392, Rockefeller Archive Center (hereafter RAC).

8. Memos, November 18 and 23, 1942; and Miller to William Vogt, November 24, 1942, RFA, R.G. 1.1, Series 200, Box 132, Folder 1631, RAC.

9. Dobzhansky to Miller, November 24, 1942, RFA, R.G. 1.1, Series 200, Box 132, Folder 1631, RAC.

10. Dobzhansky's Oral History Memoir, Columbia University Oral History Program, cited by William B. Provine, *Sewall Wright and Evolutionary Biology* (Chicago: The University of Chicago Press, 1986), 386–87.

11. William B. Provine, "Origins of the Genetics of Natural Population Series," in R. C. Lewontin, J. A. Moore, W. B. Provine, and B. Wallace, eds., *Dobzhansky's Genetics of Natural Populations I–XLIII* (New York: Columbia University Press, 1981), 11.

12. Ibid., 21–42.

13. Ibid., 61.

14. The paper trail of this episode is very thin. See Dobzhansky to Wright, January 26, 1943; Sewall Wright Papers, (hereafter Wright Papers), American Philosophical Society (hereafter APS); and Provine, *Sewall Wright*, 386–88.

15. Bentley Glass, ed., *The Roving Naturalist: Travel Letters of Theodosius Dobzhansky* (Philadelphia: American Philosophical Society, 1980), 81.

16. "A Fundamental Study," (see citation 7).

17. Interview of Antonio Brito da Cunha with author, Sao Paulo, March 22, 1990.

18. Dobzhansky, Oral History Memoir, transcript, 548 (November 14, 1962).

19. A. G. Lagden Cavalcanti, "Centro de Pesquisa de Genética da Faculdade Nacional de Filosofia da Universidade do Brasil," *Atas*, 53–58.

20. Theodosius Dobzhansky and Crodowaldo Pavan, "Studies on Brazilian Species of *Drosophila*," *Boletim de la Faculdade de Filosofia, Ciências e Letras da Universidade de São Paulo Biologia Geral* 4 (1943): 7–72; Crodowaldo Pavan and A. Brito da Cunha, "Espécies Brasileiras de *Drosophila*," *Boletim de la Faculdade de Filosofia, Ciências e Letras da Universidade de São Paulo Biologia Geral* 7 (1947): 20–64.

21. Dobzhansky, Oral History Memoir, 559.

22. Ibid., 556.

23. Following Brito da Cunha, interview with author, these eight were the key players. Dobzhansky, in addition to these, usually included his wife, Natasha, and E. Nascimento Pereira. Pavan listed thirteen original members in a 1953 letter and the following year put the number at eleven; see Pavan to Rubens Maciel, February 24, 1953; Pavan to Miller, June 4, 1954; and "Research Project of Study of Population Genetics and Ecology of Tropical Organism," RFA, R.G. 1.2, Series 305, Box 45, Folder 390, RAC. Others, such as Dreyfus, Frota-Pessoa, and Freire-Maia were clearly involved.

24. This description of the 1948–49 results is from Brito da Cunha, interview with author.

25. Burla to Miller, December 1, 1949, RFA, R.G. 1.2, Series 305, Box 45, Folder 390, RAC.

26. Theodosius Dobzhansky, "Genetic Structure of Natural Populations," *Carnegie Institution of Washington Yearbook* 48 (1948–49): 201–12.

27. Ibid.; see the map reproduced on page 204.

28. Ibid., 206.

29. H. Burla, A. B. da Cunha, A. G. L. Cavalcanti, Th. Dobzhansky, and C. Pavan, "Population Density and Dispersal Rates in Brazilian *Drosophila willistoni*," *Evolution* 31 (1950): 352–404.

30. Dobzhansky, "Genetic Structure," 207, referring to research eventually published as Th. Dobzhansky and A. B. da Cunha, "Differentiation of Nutritional Preferences in Brazilian Species of *Drosophila*," *Ecology* 36 (1955): 34–39.

31. H. Burla et al., "Population Density."

32. A. B. da Cunha, H. Burla, and Th. Dobzhansky, "Adaptive Chromosomal Polymorphism in *Drosophila willistoni*," *Evolution* 4 (1950): 212–35.

33. Dobzhansky to Wright, September 28, 1948, Wright Papers, APS.

34. B. da Cunha, Burla, and Dobzhansky, "Adaptive Chromosomal Polymorphism," 234; and Dobzhansky, "Genetic Structure," 210.

35. A. Brito da Cunha and Theodosius Dobzhansky, "A Further Study of Chromosomal Polymorphism in *Drosophila willistoni*, and Its Relation to the Environment," *Evolution* 8 (1954): 119–34.

36. Theodosius Dobzhansky, "Evolution in the Tropics," *American Scientist* 38 (1950): 209–21, on p. 217. This article is a foundational document of tropical ecology.

37. Theodosius Dobzhansky, "What is an Adaptive Trait?" *American Naturalist* 90 (1956): 347.

38. Sewall Wright, *Evolution and the Genetics of Populations*, 4 vols. (Chicago: University of Chicago Press, 1968–78). See 1: 125, citing Pavan et al., "Concealed Genic Variability in Brazilian Populations of *Drosophila willistoni*," *Genetics* 36 (1951): 13–30; 4: 69–70, citing Burla et al., "Population Density"; and 4: 124–25, citing T. Dobzhansky, H. Burla, and A. B. da Cunha, "A Comparative Study of Chromosomal Polymorphism in Sibling Species of the *willistoni* group of *Drosophila*," *American Naturalist* 84 (1950): 229–44.

39. B. da Cunha, Burla, and Dobzhansky, "Adaptive Chromosomal Polymorphism."

40. Brito da Cunha and Dobzhansky, "A Further Study."

41. Dobzhansky, Oral History Memoir, 576.

42. Glass, *The Roving Naturalist*, 165. There was another, chilling, suitability of the Angra dos Reis site for a particular kind of experiment. Miller noted in his diary with regard to Pavan's request for cobalt-60 to treat laboratory populations destined for the project: "Dobzhansky feels that this is probably the only place in the world where such studies could be carried out under such favorable conditions. Results are expected to throw light on what would happen in the human population of a medium sized city struck by an atom bomb." Miller, Diary, April 23, 1956, RFA, R.G. 1.2, Series 305, Box 45, Folder 392, RAC. Pavan, in a proposal sent to Miller on December 30, 1957, makes clear that the Atomic Energy Commission was interested in radiation effects on populations, e.g., of mice and *Drosophila*, to see

what might happen with human populations. The nuclear angle did not, however, play any significant role in the research or its rationale. RFA, R.G. 1.2, Series 305, Box 45, Folder 396, RAC.

43. Dobzhansky, Oral History Memoir, 576.

44. Francisco M. Salzano, interview with author, Porto Alegre, March 30, 1990.

45. Brito da Cunha and Crodowaldo Pavan report (annex #1), RFA, R.G. 1.2, Series 305, Box 45, Folder 396, RAC.

46. Quoted by Pavan in *Atas*, 63.

47. Ibid., 64.

48. Crodowaldo Pavan, "Algumas Questoes sobre o Desenvolvimento da Ciência no Brasil e as Liçoes na Area da Genética," *Ciência e Cultura* 41 (1989): 454.

49. Quoted, as the origin of the research values of Brazilian geneticists, by A. Brito da Cunha, "O Poder da Genética Classica," *Ciência e Cultura* 41 (1989): 450.

50. "Research Project of Study of Population Genetics and Ecology of Tropical Organism," 2, RFA, R.G. 1.2, Series 305, Box 45, Folder 390, RAC.

51. Antonio Rodrigues Cordeiro, "Genética no Brasil: Pasado e Futuro," *Ciência e Cultura* 41 (1989): 445. Salzano confirmed this fact: "Freire did not like *Drosophila*. Freire would say why not work with domestic flies (*melanogaster*) and Dobzhansky would get angry. Newton said inversions were not adaptive." According to Salzano, Kerr was another who did not like *Drosophila*; Salzano, interview with author. In 1951 Freire founded the Laboratory of Human Genetics at the University of Paraná, Curitiba, where, nevertheless, he wrote an important series of papers on chromosomal polymorphism, a favorite topic of Dobzhansky, in eight natural populations of *Drosophila*. See Newton Freire-Maia, "Laboratorio de Genética Humana da Universidade do Paraná," *Atas*, 213–17.

52. Cordeiro, "Genética no Brasil," 446.

53. A. Brito da Cunha, *Perspectivas Atuais e Futuras das Ciências Biologicas no Brasil* (Brasilia: CNPq, 1990), 11. For other recollections of Miller, see A. Brito da Cunha, "Homenagem ao Dr. Harry M. Miller Jr.," *Boletim da Sociedade Brasileira de Genética* 7 (1965): 2–3; Cordeiro, "Genética no Brasil," 445–46; *Historia da Ciência no Brasil: Acervo de Depoimentos* (Rio de Janeiro, 1984), 79–81.

54. Dreyfus to Edgard Santos, February 16, 1948, RFA, R.G. 1.2, Series 305, Box 45, Folder 389, RAC.

55. But as Nancy Stepan has pointed out to me, the vogue of eugenics could have also worked the other way, to encourage Mendelism as a way of escaping the aura of pseudoscience surrounding eugenics.

Contributors

JOSEPH COTTER is a doctoral candidate in the Department of History at the University of California, Santa Barbara.

MARCOS CUETO is a researcher at the Instituto de Estudios Peruanos of Lima, Peru. His articles on Latin American science and medicine have been published in *Isis, The Bulletin of the History of Medicine, The Hispanic American Historical Review,* and other journals.

DEBORAH FITZGERALD is Associate Professor of the history of technology in MIT's Program in Science, Technology, and Society. She is author of *The Business of Breeding: Hybrid Corn in Illinois.*

THOMAS F. GLICK is Professor of History at Boston University and author of *The Comparative Reception of Darwinism, The Comparative Reception of Relativity, Einstein in Spain,* and the chapter on science in Volume VIII of the *Cambridge History of Latin America.*

ARMANDO SOLORZANO is Assistant Professor at the Program of Family and Consumer Studies of the University of Utah and author of *The Rockefeller Foundation in Mexico: Nationalism, Public Health and Yellow Fever, 1911–1924.*

STEVEN C. WILLIAMS is a doctoral candidate in the Department of History at the University of California, Los Angeles.

Index

Acosta, Ricardo, 80, 81
Adams Act (1906), 76
Aedes aegypti (mosquito). *See* Yellow fever
Agudelo, Saul Franco, xviii*n*2
Alemán, Miguel, 80, 81, 102, 104, 106, 109, 119*n*59
Alvarado, Salvador, 61, 62
Alves, Tomas, 40, 42
Animal diseases, 106
Antunes, Waldemar, 32
Arbona, Guillermo, xvii*n*1
Argentina: medical education in, 8, 9, 11–13, 21*n*36, 21*n*39, 22*n*50, 131, 132; physiology research in, 132–34, 142; Rockefeller Foundation in, x, xi, 4, 5, 127–28; and yellow fever in Brazil, 41, 50*n*83
Arnove, Robert, xviii*n*2
Ashford, Bailey K., 20*n*17, 20*n*19
Aste-Salazar, Humberto, 130
Avila Camacho, Manuel, 97, 102, 135; agrarian policy of, 105, 108, 120*n*67; and the Mexican Agricultural Project, 103

Bailey, Joseph C., 92*n*6
Barbosa, Placido, 27, 47*n*6
Barreto, Antonio Luis de Barros, 35, 37, 43, 51*n*106
Barroso, Sebastião, 31–32
Barrus, Mortimer F., 85
Batista Luzardo, Joao, 50*n*77
Beans, 106, 107, 123*n*103
Benton, Louisa Barclay, 136, 142, 147*n*44
Berliner, Howard S., 21*n*40
Berman, Edward H., xviii*n*2, 68*n*1
Bethell, Leslie, 19*n*11
Birch, Carrol L., 48*n*30
Bolivia, xi, 5
Boll weevil, 74, 75
Bonner, Thomas Neville, 21*n*40, 22*n*56
Borlaug, Norman E., 77, 78, 82
Bradfield, Richard, 77
Brazil: and Britain, 49*n*66; diseases in, 47*n*4; genetics research in, xvi, 149–50, 159–61; hospitals in, 11, 21*n*43; impact of federalism on public health, 23, 25, 31, 45–46;

medical education in, 3–4, 8, 12, 16, 20*n*18, 20*n*25, 21*n*35, 21*n*46, 39, 127, 131, 132, 143; physiology research in, 136–38, 142, 147*n*64; the "Prestes Column" insurrection in, 32–33; public health response to yellow fever, 23–26, 28–32, 37–44; Rockefeller Foundation in, x, xi, xii, xv, xvi, 3–4, 5, 6, 12, 14, 17, 19*n*16, 21*n*46, 27–38, 134; U.S. relations with, 3, 19*n*15; waterworks projects in, 35–36, 49*n*54. *See also* Argentina: and yellow fever in Brazil; National Department of Public Health (DNSP); Sociedades de Beneficencia; Vaccine Riots of 1904; Yellow fever
Brett, Homer, 48*n*34
Brieger, Friedrich, 150, 151
Britain, 2, 19*n*5, 49*n*66
British Guiana, 19*n*6
British West Indies, 2
Brown, E. Richard, xviii*n*2, 92*n*5
Brown, Theodore M., 21*n*33
Bubonic plague, 3, 50*n*77, 54
Bullock, Mary Brown, 15, 22*n*60
Burla, Hans, 154–58

Campos, Franklin Augusto de Moura, 130
Cangi, Ellen Corwin, 21*n*41
Cannon, Walter B., 130, 135, 140–42, 146*n*22, 147*n*44, 148*n*82
Cárdenas, Lázaro, 97, 99, 101, 108, 135
Carnegie Foundation, 15
Carnival, 34, 49*n*47
Carranza, Venustiano, 53, 54, 61, 99
Carrillo Puerto, Felipe, 61–62, 67, 70*n*40
Carson, Hampton, 159
Carter, Henry Rose, 30, 48*n*28
de Castro Santos, Luiz Antonio, 20*n*23, 45–46, 47*n*5, 48*n*26
Cavalcanti, A. G., 154–56, 159
CCEFA (Cruzada de Cooperaçao na Extinçao da Febre Amarella) (Brazil), 44, 51*n*100
Central America, xii, 2, 19*n*7
Ceylon, 19*n*5
Chagas, Carlos, 37

166

		DATE DUE	